Best Wishes
Austin R. Wilkins
February, 1978

Library of
Davidson College

TEN MILLION ACRES OF TIMBER

Austin H. Wilkins

TEN MILLION ACRES OF TIMBER

The Remarkable Story of Forest Protection
in the
Maine Forestry District (1909–1972)

by

Austin H. Wilkins

Former Maine Forest Commissioner (1958–1972)

TBW Books, Publisher, Woolwich, Maine 04579

© 1978 by Austin H. Wilkins
All rights reserved.
Except for purposes of review, no part of this book may be reproduced in any fashion whatsoever without the consent of the author and the publisher.
Library of Congress Catalog Card Number 78-60181
Published by Thea Wheelwright, TBW Books
Box 58, Day's Ferry Road, Woolwich, Maine 04579
ISBN 0-931474-02-7 Hardcover edition
ISBN 0-931474-03-5 Softcover edition
Composition by Pine Tree Composition, Lewiston, Maine
Manufactured in the U.S.A. by Halliday Lithograph Corporation, Quincy, Mass.

Illustrations Reproduced Courtesy of the Following:

Great Northern Paper Co., Millinocket, Me., P.O.W. Compound, Seboomook, pp. 182–83; Fire Prevention posters, 232–33
Donald E. Johnson, Westbrook, Maine, Mt. Katahdin, p. xxiv
Maine State Library, Augusta, Me., Lottery Ticket, p. 12
Francis Moore, Grand Lake Stream, Me., P.O.W. Princeton pictures, pp. 180–81, 183
Pingree Timberlands, Bangor, Me., David Pingree, p. 3
Scott Paper Co., Winslow, Me., Typical Pulpwood Camp, p. 226
James W. Sewall Co., Old Town, Me., Flagstaff and Millinocket CCC photos, p. 190
State Inland Fisheries and Wildlife Department, Tragedy in the Woods, p. 118
Webber Timberlands, Bangor, Me., John P. Webber, p. 3

For pictures and maps not credited to the above, Maine Forest Service

DEDICATION

This book is dedicated to all members of the former Maine Forestry District Advisory Committee with many of whom the author was privileged to be closely associated for many years:

Blaisdell, George M. — International Paper Co.*
Bork, John H. — Brown Co., N.H.
Bradford, Grover — Pingree Timberlands
Burns, Kenneth — S. D. Warren Paper Co.
Carlisle, George D. — Prentiss & Carlisle Co.
Clement, Phillip P. — Prentiss & Carlisle Co.
Crocker, Floyd M. — St. Regis Paper Co.
Currier, Ralph — Eastern Pulpwood Co.
Demeritt, Dwight B. — Dead River Co.
Eggelston, William — Eastern Corporation
Ellis, Donald — Scott Paper Co.
Freedman, Louis J. — Penobscot Development Co.
Garland, Buhrman B. — Scott Paper Co.
Giddings, Edwin L. — Penobscot Development Co.
Hartranft, John L. — Oxford Paper Co.
Herr, Clarence S. — Brown Company, N.H.
Hilton, William — Great Northern Paper Co.
Kugelman, Lawrence — International Paper Co.
LaBonta, Robert R. — Scott Paper Co.
Maines, John T. — Great Northern Paper Co.

Melcher, Edward — S. D. Warren Paper Co.
Merrill, Robert W. — Penobscot Development Co.
Mitchell, Roger J. — Georgia Pacific Corp.
Pearson, Frank — Eastern Pulpwood Co.
Philbrick, William — Coburn Heirs
Pierce, James M. — Madigan & Pierce
Sawyer, Omar A. — Hollingsworth & Whitney
Sawyer, George C. — Dunn Timberlands
Semonite, David — J. M. Huber Corp.
Sleight, Charles K. — J. M. Huber Corp.
Sinclair, John G. — Seven Islands Land Co.
Stedman, Arthur F. — Scott Paper Co.
Wheatland, Stephen — Pingree Timberlands
Weller, Herbert J. — St. Regis Paper Co.
Williams, Niles C. — Dead River Co.
Wellman, Bradford S. — Seven Islands Land Co.
Weiland, George W. — Dead River Co.
Wing, Morris R. — International Paper Co.
Wood, Raymond J. — Diamond International Corp.

* Company affiliation at time of appointment between the years 1948 to 1972

Seven Islands Land Company

P.O. BOX 116 · 15 COLUMBIA STREET · BANGOR, MAINE 04401
TELEPHONE (207) 945-3022

Early settlers in North America found abundant resources they could use without concern. Wildlife and forests appeared endless. Forests, especially to settlers interested in homestead farming, were simply in the way. They had to be cleared for settlements and farming.

Often the clearing was done by fire - a match was cheaper and easier than axes and hoes. The effects of such fires were both beneficial and devastating. Over time, soil damage and loss of useable wood presented both immediate and long-term problems to present and future citizens.

Maine, too, experienced this hazard of settlements, but fortunately, not to the extent of neighboring states and provinces.

Fortunately for this and future generations, much of the forest land in Maine has been held in private ownership by people with good foresight. They early established the policy of forest protection to fight and prevent such fires. We can, therefore, look to the future with abundant forest resources because of their conservation and protection policies.

The history of the Maine Forestry District is the account of these policies. The story has important background and lessons for resource managers and users today.

Retired Forest Commissioner Austin Wilkins recounts this story in "Ten Million Acres of Timber." He describes the dedication and hard work of those connected with the resource - land owners, government personnel and others.

His well-researched and documented account passes the message on to the future generations in a form which clearly points out good citizenship and the true meaning of meeting responsibilities for land and forest resource stewardship.

Being able to spend most of my working years involved in this protection and management work with Austin and many of the people he mentions has been an indescribable privilege.

John G. Sinclair, President
Seven Islands Land Company

INTERNATIONAL PAPER COMPANY
JAY, MAINE 04239, PHONE 207 897-3474

WOODLANDS & WOOD PRODUCTS OPERATIONS—REGION VI

ROBERT R. EASTMENT, Manager

      Trees, forests and wood fibre have been a way of life for the people of Maine for over 3 1/2 centuries. In the early period of growth and development, Maine was the largest producer of sawn lumber in the United States. In 1977, Maine once again emerged as the No. 1 producer of paper in the nation. It is truly remarkable that Maine's forests, comprising nearly 90% of the land area, could have sustained such growth and annual wood production throughout centuries of use.

      Over the course of time, a gradual change has occurred in the use of these forests; from the original single purpose for wood fibre, to the present day multiple use to include all of the other resource values. It has occurred to me that the people of Maine, who use these vast forests for their very livelihood, owe much to those farsighted landowners who, long ago, took the necessary legislative action to protect this great renewable resource.

      About 10 million acres of this timberland (nearly 1/2 of the land area of the State) falls within the unorganized townships and plantations which were afforded forest protection from 1909 to 1972 by the Maine Forestry District. This protective District was created as the result of voluntary action of the paper companies and private landowners who, by self-imposed taxes, financed the forest protection efforts of the Maine Forest Service within these so-called Wildlands. After 63 years, many still ponder the fact that this protection system worked so successfully.

      In 1972, a legislative act established the Department of Conservation and the Maine Forestry District was phased out as a separate entity. At that time, it seemed appropriate that someone should call attention to and record this remarkably history of the Maine Forestry District as, surely, there has existed nothing like it before. By coincidence, Austin Wilkins' retirement from State service as Commissioner of the Maine Forest Service also occurred in 1972. Here was a tremendously dedicated man whose career of 44 1/2 years with the Forest Service actually spanned nearly 2/3 of the entire life of the Maine Forestry District. Who could possibly tell the story better than Austin?

      As a representative of the Pulp & Paper Industry, I express my heartfelt thanks to that great public servant, Austin Wilkins, for his deep personal interest in writing this book. We also want to acknowledge with admiration and appreciation the service of previous Forest Commissioners, Supervisors, Chief Fire Wardens, Patrolmen, Tower Men, Rangers, Pilots, their wives, and others in the Forest Service, whose love for the forests caused them to serve far above and beyond the call of duty to protect this great natural heritage for those who follow.

      We owe them much.

Morris R. Wing
Morris R. Wing
Manager - Maine Woodlands

ACKNOWLEDGMENTS

Obviously the preparation of this documentary history involved contacts with many people. Without their help much of the written material would not have been possible.

I wish to express my gratitude to the Maine Forestry District Advisory Committee, with whom I was privileged to be closely associated for many years, for their excellent cooperation on a forest commissioner-landowner basis. It was a bond of trust and mutual confidence.

My thanks also go to the State Forestry Department for the use of library facilities, the privilege of taking out material for review, and for contributed stenographic help and reproduction of many photos, tables, charts and other useful material. Similar appreciation goes to the Paper Industry Information Office for typing and copying.

There are a number of individuals who should be singled out for special mention:

My particular thanks go to Morris Wing, of International Paper Company, and John Sinclair, of Seven Islands Land Company, whose encouragement inspired me to write this history. It has been a most enjoyable and rewarding experience.

As the format of this book began to take shape, some valuable assistance came from Ben Pike, who served as a sort of special consultant. His general over-view resulted in helpful and timely suggestions.

A deep sense of gratitude goes to my former department associates Fred Holt, Robley Nash, Richard Sawyer, Henry Trial, Russell Cram, Walter Gooley, Robert Pendleton, William Cross, former Forest Commissioner A. D. Nutting and Arthur Stedman, of Scott Paper Company, for their valuable assistance.

I wish especially to acknowledge the very valuable help and patience of my former secretary Anna Stanley, who, in addition to her continuing secretarial duties to my successor, managed to type a first-draft copy of this M.F.D. history from long-hand writings, notes and tabulations.

Also my special thanks go to Robert I. Ashman, Professor Emeritus of Forestry, former head of the School of Forest Resources, University of Maine Orono, Maine, for his most helpful review of the original written material.

Finally, my deep appreciation goes to Walter Macdougall, who undertook the preliminary task of organizing the manuscript material and making necessary alterations in preparing it for the publishers.

August H. Wilkins

CONTENTS

Introduction		xix
"Firsts" in Forest Protection		xxi
I.	The Forest and Land Ownership	1
II.	The Threat and the Concern	25
III.	Creation of the M.F.D.	39
IV.	Organization and Warden Service	49
V.	Financing and Accounting	61
VI.	A Testimony to Cooperation	79
VII.	Watchmen and Telephone Lines	95
VIII.	Fire Suppression and Reporting	121
IX.	Radios and Aircraft	147
X.	War Times and the C.C.C.	171
XI.	The Spruce Budworm Program	195
XII.	Personnel and Public Relations	209
	Appendices	237
	Bibliography	309

ILLUSTRATIONS

Austin H. Wilkins	frontis.
Unusual aerial photo of Mt. Katahdin	xxiv
David Pingree and John P. Webber	3
Collection of Surveyors' private marks	6
One of two remaining unsold lottery tickets	12
A land agent's timber-cutting grant	20
Photo of a lightning hit on a white pine stub	24
The fury of a fire	28
The first and last forest commissioners	38
Home of the Forestry Department, Augusta	60
Compact Signing Ceremony, Quebec	84
Compact Signing Ceremony, Fredericton	85
Pact documents	86, 88–90
Old Wooden Tower, Depot Mountain, 1909	94
Monument to first lookout towers	99
William Hilton, first lookout, and others	99
Close-up of tablet inscription	100
Unused wooden tower, Mt. Washington and an early lookout station	101
Watchmen's camps	102
Early crude wooden lookout towers	104
Log cribbed wood towers and others	105
Some of the old solidly enclosed towers	106
Steel towers with wood cabins, built 1917	107
Results of severe weather	108
Steel tower with wooden cab, and latest all steel tower	109
Helicopter landing pilot and author on Mt. Bigelow	111
Mt. Bigelow lookout tower, erected 1917	112
Remote guardian of the forest	114
Inside typical watchman's tower cabin	114
Tragedy in the woods	118
Horse teams hauling fire equipment to a fire	122
Lobster Mountain Fire, 1911	124
Fire equipment flown in by plane and used on shoreline	127
A pile of pulpwood consumed by fire	132
Rex Gilpatrick and jeep and John Mitchell's old state vehicle	133
Bonded Canadian fire fighters	136–37
Equipment storage for quick loading	139
Canadian Pacific Railroad patrolman	140
Loading and unloading fire equipment	145
Parachute drops of equipment, now standard procedure	146
Float plane picking up water	160
Water-bombing a fire	160

One of the Stinson Planes in Moosehead Lake	163
Parachute drops of supplies to watchmen on remote towers	164
Water-bombing forest fires, an effective tool in suppression work	169
Lookout tower closed during war period	170
Personnel notification	176
German P.O.W.'s	180–81
Seboomook P.O.W. camp	182
More P.O.W.'s	183
CCC camps	190
Planes spraying in ragged formation	194
Spruce and fir killed by budworm defoliation	197
Low-flying team of spray planes	199
First annual meeting of landowners and chief wardens	208
Council order 916	218
Fiftieth Anniversary of M.F.D.	221
A warden patch and samples of badges	222
Forest fire danger class day	224
Typical pulpwood camp, Appleton township	226
Fire danger road signs	230
Industrial bilingual forest fire prevention posters	232–33

TEXT TABLES

Schedule Maine Woodlands 1878	8
Area of land classes 1971	8
Area of commercial forest land ownership 1971	9
State statistics	9
State valuation for all unorganized territories 1973	23
Unorganized territory, 1891	31
Forest area in Maine (updated changes)	45
M.F.D. acreage by classification 1971–72	45
Municipalities joining M.F.D., and gross acreage of forests	46
Fire control plans, 1951 and 1952	50
Schedule of nomenclature changes	54
Organization of M.F.D. in 1925	58
M.F.D. personnel summary 1961	59
Forest fire control division field personnel 1971	59
M.F.D. expenditures 1909	64
M.F.D. expenditures fiscal year ending 6/30/72	64
M.F.D. financial statement 6/30/72	65
M.F.D. forest fire tax assessments	66
Comparative financial statements 1909 and 1959	68
Petty cash account bank statement	73
Status of fire reserve account 1955–72	76
Bills for fire fighters' supplies	125
Fire organization buildup	128

Fire fighting wage rates	130
Equipment rates 1972	135
Railroad main track mileages	141
Radio inventory and value 1972	155
Chart showing reduction of lookout tower service	166
Hours flown by contractual air patrol	168
Schedule of Civilian Conservation Camps in M.F.D.	191
Summary of aerial spraying for budworm control	201
Budworm M.F.D. tax assessments	202
Spruce budworm project—500,000 acres	204
Reports of forest commissioners	209
M.F.D.'s accident rate 1970–71	220
M.F.D. fire danger stations	225
See Appendices Contents for other tables	237

MAPS IN TEXT

Land Survey Designations	11
Public lots	18
A 1,000 acre lot	19
Fire control map and 1972 organization chart	48
Map showing historical fire locations	120
Map showing frequency assignment locations	154
Map showing areas of fir and spruce mortality	196
Map of proposed spray areas	206

Maps Included in Cover Pocket

Early Moses Greenleaf map
Public Reserved Lands, 1976
A topographical map

INTRODUCTION

In the long history of the State of Maine Forestry Department nothing is more unique than the period covering the creation and administration of the Maine Forestry District (1909–1972). It is the fascinating story of a peoples' concern and the action taken for a better forest protection system in the unorganized territory of the state. All the more remarkable is the fact that it worked so successfully for sixty-three years.

This administrative era ended with the passage of the Maine Tree Growth Tax Law (Chapter 616, Public Law, March 10, 1971, Section 1608) by the Special Session of the Maine 104th Legislature.

Upon request I have attempted to put together a chronology of this historic period as a distinct chapter in the annals of forest protection in Maine. What follows is a continuous account of the flow of activities, events, changes, and legislation that took place, along with occasional human interest anecdotes

Obviously this historic documentary is the result of much research and access to the voluminous wealth of forest commissioners' reports, old diaries, journals, letters, museum and archive files, private company family records, and personal interviews with people associated at one time with the happenings of the Maine Forestry District.

Especially helpful was access to the most complete file of letters, papers and reports kept by the late Chief Warden John Mitchell for the period 1915–1927, which is now stored in the Augusta Office. Numerous references from these records appear in this M.F.D. history.

I am privileged to be able to write from the vantage point of forty-four years of public service with the Forestry Department, and to look back and recall many personal experiences. It is important to record this personal input before memories fade and the facts become obscure.

In no way is this an exhaustive study, but rather an attempt to recapture some of the more salient points in what occurred. To those who wish to pursue in greater detail some of the subjects covered, a special effort has been made to prepare a thorough bibliography.

It is also to be noted that the material used in this history has a cut-off date of 1972 to coincide with the termination of the Maine Forestry District. Shortly thereafter some important changes occurred, but they were not considered to be within the scope of this study. In a few instances some tables have been updated, but no extended narrative of events beyond 1972 has been added.

In the pages to follow the reader will have the opportunity to retrace the remarkable story of forest protection for ten million acres of Maine timberlands.

AUSTIN H. WILKINS

"FIRSTS" IN FOREST PROTECTION

FIRST in the country with "common and undivided" forest land ownership management of large areas – 1840

FIRST state in the country to precede federal action for forest fire protection by legislative act – 1891

FIRST in the country to establish a rate for forest fire fighting "not less than fifteen cents per hour" by legislative act – 1891

FIRST in the country to establish organized patrols by canoe on the rivers, lakes, and streams in the unorganized territory by legislative act – 1903

FIRST continuously operated forest fire lookout tower in the country – Squaw Mountain, 1905–1967

FIRST in the country for private forest landowner self-imposed forest fire protection tax (creation of M.F.D.) by legislative act – 1909

FIRST in the country with legislative action for woods closure by governor's proclamation – 1909

FIRST of three states in the country to receive $10,000 for forest fire protection under Weeks Law – 1911

FIRST to use panorama profile lookout tower maps – 1917

FIRST in the country issuing revolvers, ammunition, holsters, and handcuffs to chief forest fire wardens and lookout watchmen to protect against sabotage of Maine's wildlands – World War I – 1918–1919

FIRST in the country paying fire fighters by check or cash "right on the stump" – 1920

FIRST state conservation agency east of the Mississippi to use aircraft, especially float planes, for forest fire protection – 1927

FIRST full-time professional forest entomologist in the country, H. B. Peirson, appointed by Forest Commissioner Samuel T. Dana – 1931

FIRST state conservation agency to introduce forest insect surveys, using fire wardens and private industry foresters for making periodic collections – 1941

FIRST in the country to use mechanical Smokey Bear as a bilingual in northern Maine French-speaking pulpwood camps – 1956

FIRST in the country creating a Northeastern Forest Fire Protection Compact: authorized by Congress in 1949 and ratified by the New England States and New York in 1949–1950; authorization by Congress for Canadian participation in 1952 and joinder action by Quebec – 1969, and New Brunswick – 1970

TEN MILLION ACRES OF TIMBER

The Silver Smokey statue is inscribed: "Highest award to the Northeasters Forest Fire Protection Commission in Forest Fire Prevention, by the National Advertising Council, Inc., National Association of State Foresters, U.S. Department of Agriculture, Forest Service, July 10, 1972"

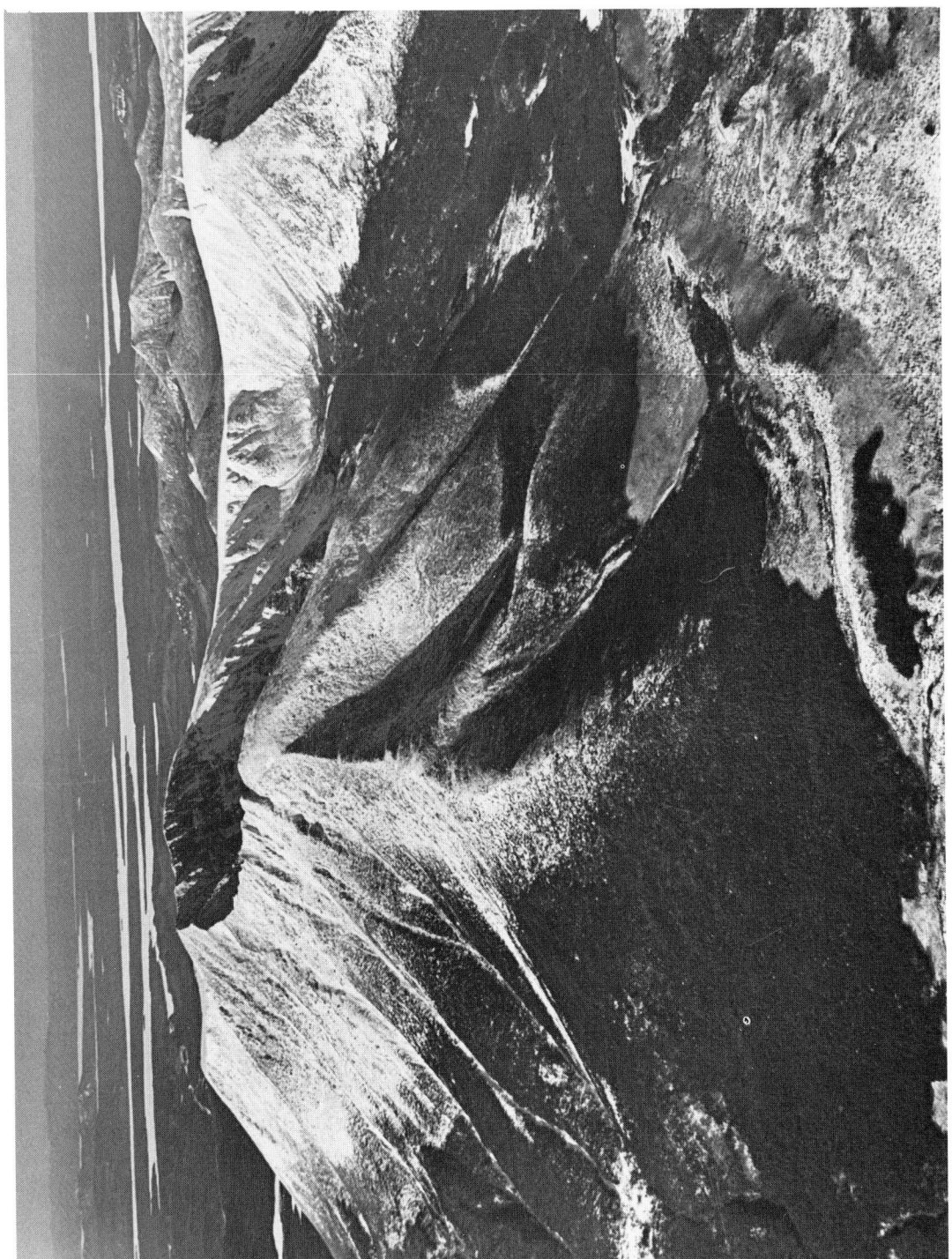

Unusual Aerial Photo of Mt. Katahdin Range

I

THE FOREST AND LAND OWNERSHIP

> *What is most striking in the Maine wilderness is the continuousness of the forest, with fewer open intervals or glades than you had imagined.**

Maine and forest, the two are synonymous in the minds of many people, and rightly so. What was true for the Indian and the pioneer remains so for us. The forest is Maine's great natural resource, standing tall and spreading over 17,748,600 acres. It is a provider, a source of a multi-million-dollar business, and a priceless heritage of growth and beauty.

In 1909 the Maine Forestry District (or the M.F.D. as it quickly became known) was created to be the guardian of this vast treasure. The story of the M.F.D. and its years of service starts, properly, with this date, but knowledge of what had transpired over prior years, the shift from public lands to private and the growing concern over preservation and conservation, is essential for an appreciation of the District's reason for being and its mission.

At the close of the Revolution, the Commonwealth of Massachusetts found herself in serious financial straits. Her treasury was empty, tax burdens heavy, credit near that of bankruptcy, paper currency worth in the market scarcely ten per cent of its normal value, commerce next to nothing, and her public indebtedness building at an alarming rate.

Faced with these conditions, Massachusetts turned her attention to her Eastern Lands (The District of Maine) where she owned in fee simple vast areas of public domain. Since these lands were considered of some value, their sale seemed to promise a much needed source for

---

* Thoreau, *The Maine Woods* (1853).

new revenue. The land was originally recognized for its value as a place for new settlement and development, but in later years, the emphasis shifted to timber. The report of the land agent to the Legislature in 1848 contains this prophetic statement, "The value of the land consists entirely in its timber and generations to come will not furnish a demand for it for any other purpose."

By 1835, one of the greatest land-boom speculations in the history of the country was occurring in the disposition of the public domain in Maine. Forty-three years later, the Maine land agent reported that all of the public lands of the state had been disposed of and none remained that the state could grant for homestead settlers.

Large blocks of timberland were first purchased by shrewd people for speculation with the hope of a resale for profit. Some became proficient traders in timberlands on a "buy cheap and sell dear" basis. As early anticipations of homesteading, the making of farms and villages, the building of industries and roads, gave way to reality, the "resale for profit" owners found it more to their advantage to hold their lands for the profit derived from stumpage sales. A good source of income came from the sawmills "downriver," when Maine became the lumber capital of the world. But by 1909 the sawmill era had reached its peak.

Its passing ushered in a new era of purchase of large tracts of timberland for the pulp and paper industry. It was quickly recognized that to assure a continuous supply of raw materials, it would be necessary for corporations to own timberland rather than to depend upon purchase of wood from private owners. The transition from public domain to private was followed by a growing program of land acquisition that transferred titles back from private ownership to large corporate holdings, a process which is still going on.

According to the Timber Resources of Maine Report for 1971, ninety-eight per cent of the state's total acreage was in private and corporate ownership, with an estimated 100,000 owners ranging from farmers with ten-acre woodlots to corporations owning over 2,250,000 acres.

During this three hundred and fifty years of forest land transfer, two small parcels of public domain escaped and still remain in state ownership. They were not uncovered until 1930 during research for other information. These are 1,053 acres in Sheridan (now Ashland) and 216 acres in New Sweden, a total of 1,269 acres.

With the creation of corporate holdings of the pulp and paper industry, a great number of once prominent and familiar names rapidly disappeared. Not more than three or four million acres remain of what once were the holdings of men and families who shaped the course of events and policies up to and during the formation of the M.F.D. Below is given a list of a number of these timberland owners taken from

David Pingree 1841–1932
Early major owner and prudent manager who put together Pingree Timberlands, which remain today as the largest of private family forest land ownerships

John P. Webber 1831–1911
Founder responsible for early purchases of properties to form Webber Timberlands, which remain today as the second largest private family forest land ownerships

the State Assessor's Report of 1906. Their names deserve a place in any historical account of the development of Maine's forest industry.

AROOSTOOK COUNTY

Bass, J. P.
Benson, A. W.
Blake, S. H. estate
Burleigh, A. A. and E. C.
Coe, Dr. T. U.*
Donworth, J. P.
Dunn, George B., and estate of E. G. Dunn
Dwinel, Lester
Eaton, Henry F. & Sons
Griswold, Harriet S.
Hayford, William B.
Hinckley heirs and Frank Hinckley
Hunt, Frank W. & Co.
Jenness Land Company
Lord, Charles V.
Madigan, J. B. and A. W., and C. H. Pierce
Mansur, Rufus estate
Oak, John M. and Charles E.
Pingree, David*
Powers, Frederick A. and Llewellyn
Prentiss, H. E. and S. R. estates
Sawyer, Louise J.
Small, Isaac S.
Stetson, E. I., George, and Charles estates
Strickland, P. A. and S. P.
Thomas, W. W., Jr.
Webber, J. P. and C. P.

---

* The names Coe and Pingree often appear as joint owners.

HANCOCK AND WASHINGTON COUNTIES

Baldwin, Thomas W. estate
Burrall, George E.
Butterfield, Jerome
Campbell, A. & Co. and G. R. Campbell Co.
Coffin, Joseph A.
Giles, J. T.
James Murchie Sons Co.
Loggie, A. & R., of New Brunswick
Murch, E. J.
Nash, J. W. M. and F. C.
Sullivan, Cornelius
Whitcomb, Haynes & Whitney

FRANKLIN AND SOMERSET COUNTIES

Appleton, F. H.
Boston, Lanigan and Haines
Boynton Land & Lumber Co.
Bradley, Minnie A. and Sarah J.
Coburn heirs
Franklin & Somerset Land and Lumber Co.
Gray, Joshua & Sons
Haynes, J. Manchester
Lawrence Bros.
Philbrick, S. W.
Skinner, French & Co.
Underwood, G. F.
Viles & Goodwin

PENOBSCOT & PISCATAQUIS COUNTIES

Adams, S. and J. estate
Bailey, Tabor
Boynton Land Co.
Bradbury, H. W. and Eliza A. estate
Bradley Land Co.
Cassidy, John
Eastern Land Co.
Engel, William and Lumbert
Fish River Lumber Co.
Giddings, Moses
Godfrey, Abbie P.
Hallett, Francis P.
Hersey, S. F. estate
Holyoke, Caleb and F. H.
McNulty, James
Moosehead Investment Co.
Mullen, Charles W.
Pierce & Townsend
Piscataquis Iron Works
Ross, John
Smith, J. Hopkins

As these holdings passed to the paper and pulp industry, the private interests were managed by individuals who were not necessarily owners of large woodlands. During the period between 1940–1960 approximately, the following men within the industry represented the interests of substantially large woodland owners:

Bearce, George — St. Regis Paper Co.
Blaisdell, George — International Paper Co.
Buck, Hosea — Pingree Timberlands
Carlisle, George Sr. — Prentiss and Carlisle
Crocker, Floyd — St. Regis Paper Co.
Demeritt, Dwight B. — Dead River Co.
Eggelston, William — Eastern Corporation
Freedman, Louis — Penobscot Development Co.

Hendricks, Roy	International Paper Co.
Hilton, William	Great Northern Paper Co.
Kugelman, Lawrence	International Paper Co.
Madden, James L.	Scott Paper Co.
Pearson, Frank	Eastern Pulpwood Co.
Pierce, James*	Madigan and Pierce
Sawyer, Omar	Hollingsworth and Whitney Paper Co.
Sewall, James W.	J. W. Sewall Co.
Wheatland, Stephen*	Pingree Timberlands

In the period between 1960–1970, major private interests were managed by:

Bork, John	Brown Co.
Carlisle, George	Prentiss & Carlisle
Currier, Ralph	Great Northern-Nekoosa Co.
Hartranft, John	Oxford Paper Co.
Maines, John	Great Northern-Nekoosa Co.
Mitchell, Roger	Georgia-Pacific
Philbrick, Wm.	Coburn Heirs
Sawyer, George	Dunn Heirs
Semonite, David	J. M. Huber Corp.
Sinclair, John	Seven Islands Land Co.
Sleight, Charles	J. M. Huber Corp.
Stedman, Arthur	Scott Paper Co.
Weller, H. J.	St. Regis Paper Co.
Williams, Niles	Dead River Co.
Wing, Morris	International Paper Co.
Wood, Raymond	Diamond International Corp.

Some of these men served on the M.F.D. advisory committee first created in 1948. And a number of them are still members of the industry as of this writing (1973).

With the opening of Massachusetts' Eastern Lands for settlement and speculation came the need for large-scale surveying projects and the establishment of boundaries. More than one hundred years before the M.F.D. assumed responsibility for the safety of large blocks of forest, the running of lines had begun through this wilderness and demarcations so important then and now were being established.

In 1783, Massachusetts, taking positive measures in the disposition of her public lands in the District of Maine, created a land office and appointed a land agent. His primary function was to initiate the surveying of large tracts of land in the vast unorganized territory that it might be opened for sale.

---

* Exceptions—these are also large woodland owners in their own right.

Collection of Land Surveyors Private Marks or Seals

Lore Alfred	1846	
Daniel Barker	1859	
Charles Vernon Barker	1875	
Noah Barker	1860	
T. W. Baldwin		
H. W. Briggs		
C. D. Bryant		
S. T. Buzzell	1890	
T. B. Buzzell		
Turner Buswell	1875-1900	
John H. Burleigh	1900	
Zebulon Bradley	1833	
Ed. W. Bateman		N. H.
Eleazer Coburn	1820	Skowhegan, Maine
Elmer Crowley	1910	Greenville, "
Walter Craig	1914	Greenville, "
James Conners	1910	Old Town, "
Forrest H. Colby	1910	Bingham, "
Henry Crowell	1914	Skowhegan, "
Augustus M. Carter	1870	Bethel, "
Sewall Carter	1900	Brownville, "
Ezekiel L. Chase	1900	Brownville, "
Charles E. Cobb	1900	Patten, "
H. Daggett		
Lucius P. Dudley	1880	Kingfield, "
William Dwelly, Jr.	1848	
Ira D. Eastman	1899	Old Town, "
F. J. Fiske		
T. W. Furrier	1891	
P. P. Furber		
William R. Flint	1845	
R. Gilman		
W. R. Goodwin		
C. R. Goodwin		
Alex. Greenwood	1812	
Eben Greenleaf	1816	
Elmer E. Greenwood	1890	Skowhegan "
Rufus Gilmore		
E. O. Grant		Patten, "
David Haynes	1846	
Park Holland		
J. C. Hutchinson	1906	Bangor, "
J. Herrick		
Samuel Harrison		
Clifton S. Humphreys	1900	Madison, "
John Holden	1900	Topsham, "
W. H. Jenne		South Paris, "
R. Kittredge		
Joseph Kelsey	1832	
J. A. Lobley	1890	Bangor, Maine
Caleb Leavitt	1834	
F. S. Lord		N. H.
Geo. Moulton		
Andrew McIlellan		
William Monroe		Brownville, Maine
Amaziah D. Murray	1880	The Forks, "
R. E. Mullaney	1908	Bangor, "
Roy L. Marston	1910	Skowhegan, "
McKechnie—also		
Neal & McKechnie	1814	
E. McCort Macy	1904	
T. C. Norriss		
J. C. Norris	1820	
L. A. Nason		
R. M. Nason	1891	Bangor, "
Henry Nelson		Rumford, "
John Neal	1814	
(also Neal & McKechnie)		
William P. Oakes	1870	Foxcroft, "
Louis Oakes	1900	Greenville, "
William P. Parrot	1840	
Silas Peaslee	1890	Upton, "
John Pierce	1847	Solon, "
John F. Phillipi	1913	Bangor, "
Joseph Patten		
Edwin Rose	1834	
H. G. Robinson	1900	
Joseph Sewall		
James W. Sewall	1900	Old Town, "
F. Snow		
Isaac S. Small	1836	
J. H. Stuart	1891	
J. Smith Spaulding	1862	
F. H. Sterling	1906	Augusta, "
George L. Smith	1900	Augusta, "
Boswell B. Tarbox		
Moses M. Thompson	1890	Bingham, "
Al S. Teer		N. H.
L. P. Thompson		
William M. Viles	1900	Flagstaff, Maine
Samuel Weston	1811	
O. A. Wadsworth		
John C. West	1900	Lisbon and N. H.
Joel Wellington		
Unknown	1899	
Unknown	1896	
Unknown	1871	

A number of these early surveyors played an important role in laying off large tracts of forest land subdivisions for sale and/or grant in the public domain

In the next ninety-five years (1783–1878), periodic surveys were made so that eventually all the public domain was subdivided into six-mile-square wildland townships by range and number, and each block of townships was identified by one of fifteen land survey designations.

Massachusetts was not alone in this effort to survey and sell. When Maine became a state in 1820, she also created a land office and adopted the same policy of wildlife disposition as her sister state. With Maine and Massachusetts acting jointly at times and taking separate action at other times, it is little wonder that the sale of public lands in Maine, along with the establishment of titles and descriptions, became complicated.

In laying off large tracts of land and subdividing them into townships, the early surveyors inadvertently caused some interesting irregularities. It would be wrong to assume that they were careless or dishonest in their field work. There were difficulties in traveling over blow-downs, old "burns," and rugged and rocky terrain. Under these conditions, it was common practice to give liberal measurements. The running of some lines was even omitted where the going was particularly rough. Crude instruments and the lack of precise surveying methods of today's standards in turning angles and measuring distances were also factors that must be taken into consideration.

Thus it is understandable why a number of townships are not exactly six miles square (23,040 acres) and why some irregular shaped parcels of land appear upon our present maps. A glance at a minor civil division map of Maine also shows such irregular tracts as Misery Gore and Rockwood Strip, which are located west of Moosehead Lake, resulting from the shift in surveying practice from laying out blocks of townships running magnetic north and south to that of running by the true meridian.

In running lines surveyors spotted trees and set cedar corner posts properly scribed on one or more sides with their mark, date, lot, and township number. Corner posts also had several spotted "witness trees." Such posts marking the corners of wilderness tracts were monuments to these surveyors, who are to be admired for their courage and ability to endure the hardships of the back country, the elements, and their absence from any form of civilization for weeks at a time.

The marks of these surveyors, therefore, have a special significance apart from any artistic or legal considerations. They are testimonies of hard and conscientious work. Sewall Company of Old Town has an interesting card index file on many of the old and present surveyors with their marks, dates of survey and, especially interesting, comments on the quality of their work. The marks given here are those of surveyors from 1811 to 1914. A few of those closely associated with the

SCHEDULE OF MAINE WILDLANDS 1878

	Acres
Granted by Plymouth Council prior to 1783.....	3,785,488
Sold by Massachusetts, 1783-1853..............	6,752,987
Granted by Massachusetts, 1783-1853...........	1,686,712
Total conveyed in which Maine never had any interest.........................	12,225,187
Sold by Massachusetts and Maine in common, 1820-1853...........................	1,750,605
Sold by Maine, 1820-1878......................	3,573,323
Granted by Maine, 1820-1878...................	1,968,285
Total acreage in which Maine had an interest.............................	7,292,213
Grand total.................................	19,517,400

It will be seen that the State of Maine by herself sold and granted 5,541,608 acres or less than 30 per cent of the area of the state. Of this, 1,198,330 acres, or 20 per cent, was purchased from Massachusetts in 1853. The price was 30 and 1/3 cents per acre. The 3,573,323 acres sold by the state brought $2,014,221.66 or an average per acre of 56.4 cents. Of the 1,968,285 acres granted by the state, 700,000 acres were deeded to the European North American Railway Company to aid in the building of the railroad from Bangor to Vanceboro.

TABLE I

Area by land classes, Maine, 1971

Land class	Area	
	Thousand acres	Per cent
Forest land:		
Commercial	16,894.3	87
Productive-reserved[1]	220.7	1
Unproductive	633.6	3
Total forest land	17,748.6	90
Nonforest:		
Cropland[2]	894.2	4
Pasture[2]	98.1	1
Other[3]	1,056.2	5
Total nonforest land	2,048.5	10
Total area[4]	19,797.1	100

[1] Includes 31,200 acres in the Acadia National Park, 164,300 acres in the Baxter State Park, and 11,500 acres in the Allagash Waterway.
[2] Source: 1964 Census of Agriculture.
[3] Includes swampland, industrial and urban areas, other nonforest land, and 97,431 acres, classed as water by Forest Survey standards, but defined by the Bureau of the Census as land.
[4] Source: United States Bureau of the Census, Areas of Maine: 1960. (June 1967).

TABLE II

Area of commercial forest land,
by ownership classes, Maine, 1971

Ownership class	Area	
	Thousand acres	Per cent
National Forest	37.5	(1)
Other Federal	35.8	(1)
State[2]	163.0	1
County and municipal	75.2	1
Total public	311.5	2
Forest industry	8,255.0	49
Farmed-owned	1,122.1	7
Total	9,377.1	56
Miscellaneous private:		
Individual	6,797.2	40
Corporate	408.5	2
Total miscellaneous private	7,205.7	42
All ownerships	16,894.3	100

[1] Less than 0.5 per cent.
[2] Does not include 317,414 acres in public lots on which the timber and grass rights are privately owned.

TABLE III
State-wide Statistics

16 Counties
22 Cities
419 Towns (municipalities)
55 Plantations (43 wildland plantations organized for school purposes)
(22 organized plantations)
416 Unorganized Townships

8 Gores*
3 Surpluses
2 Points
1 Patent
2 Tracts
6 Strips
4 Grants
2 Indian Purchases
1 Indian Township
1 Peninsula
(All of these subland divisions are in the Unorganized Territory, part of the minor civil land divisions survey systems)

*Gore - This is a term unique to the New England states and especially significant in Maine as associated with early township surveys. Its origin is a dressmaker's term for a three-cornered piece of cloth sewn into a garment as a tuck or gore.
  When Massachusetts and Maine laid off large tracts of land in the unorganized territory and subdivided them into townships, many varied in size with some parcels completely left out or ignored - hence the term gore.

laying off of early large tracts of public lands within the District of Maine were Noah Barker, Zebulon Bradley, Eben Greenleaf, Park Holland (1794),* Caleb Leavitt, and J. C. Norris.

The international boundary between Maine and the Province of Quebec and New Brunswick was in dispute for over 150 years and not until as late as 1925 was there final agreement on the offshore demarcation in the Bay of Fundy. Disputes and settlement involved the Treaty of Peace, Paris 1783; Treaty of Ghent, Belgium 1831; and finally the Articles of Separation and the Webster-Ashburton Treaty of 1842.

The northeastern boundary of Maine, after a dispute of nearly sixty years, was agreed upon under the 1842 treaty. The following description is taken from the Articles of Separation.

> It is hereby agreed and declared that the line of boundary shall be as follows: Beginning at the monument of the river St. Croix as designated and agreed to by the Commissioners under the fifth article of the Treaty of 1794 between the Governments of the United States and Great Britain; thence north following the exploratory lines run and marked by the surveyors of the two Governments in the years 1817 and 1818 under the fifth article of the Treaty of Ghent to the intersection with the river St. John and to the middle of the channel thereof. . . .

As stated earlier, a system of fifteen land survey designations developed as a result of periodic surveys. These are still recognized as a basis for individual township identification. They are especially useful for purposes of title search, location, taxation, mapping, etc. Failure to associate the proper land survey designation with each township can result in confusion. The following illustrates this point. The three townships listed below bear the same range and number, but have quite different locations:

T. 3 R. 4 W.E.L.S. (West of the East Line of State)
T. 3 R. 4 N.B.K.P. (North of Bingham Kennebec Purchase)
T. 3 R. 4 B.K.P.W.K.R. (Bingham Kennebec Purchase, West of Kennebec River)

Because of their special significance both historically and from the standpoint of present utility, an outline of the fifteen survey designations is given below. It will be helpful to refer to the Land Survey Designation Map and especially to the 1829 Greenleaf map while reading this section.

Each of the Bingham land purchases has its own survey designation. There is considerable history concerning these lands, involving the colorful figures of General Henry Knox and Colonel John Black, but omitted here since the subject at hand deals primarily with the

---

* Apparently David Hanes assumed Park Holland's mark in 1846.

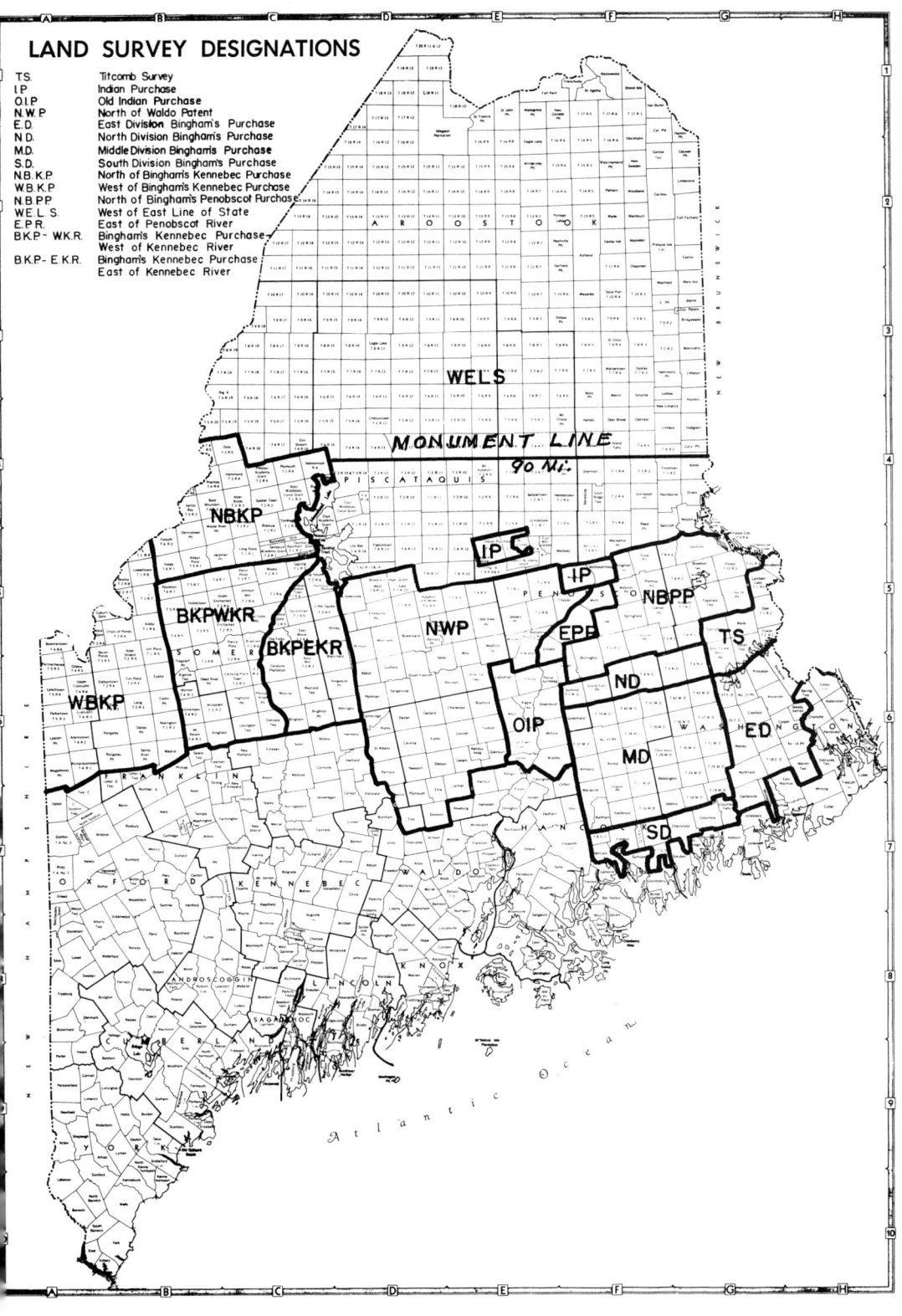

LAND SURVEY DESIGNATIONS

T.S.	Titcomb Survey
I.P.	Indian Purchase
O.I.P.	Old Indian Purchase
N.W.P.	North of Waldo Patent
E.D.	East Division Bingham's Purchase
N.D.	North Division Bingham's Purchase
M.D.	Middle Division Bingham's Purchase
S.D.	South Division Bingham's Purchase
N.B.K.P.	North of Bingham's Kennebec Purchase
W.B.K.P.	West of Bingham's Kennebec Purchase
N.B.P.P.	North of Bingham's Penobscot Purchase
W.E.L.S.	West of East Line of State
E.P.R.	East of Penobscot River
B.K.P.-W.K.R.	Bingham's Kennebec Purchase – West of Kennebec River
B.K.P.-E.K.R.	Bingham's Kennebec Purchase – East of Kennebec River

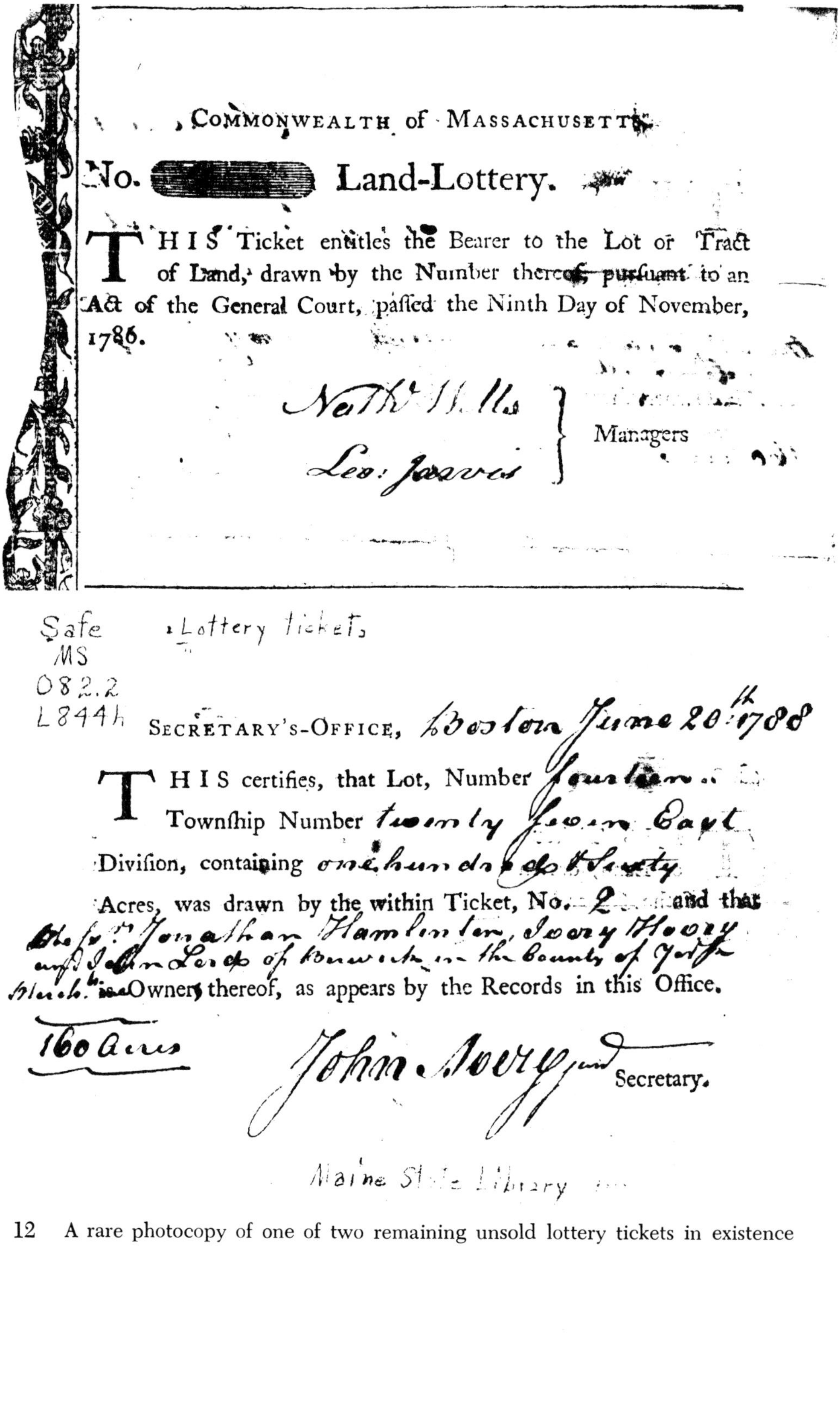

COMMONWEALTH of MASSACHUSETTS

No. ▬▬▬▬ Land-Lottery.

THIS Ticket entitles the Bearer to the Lot or Tract of Land, drawn by the Number thereof, pursuant to an Act of the General Court, passed the Ninth Day of November, 1786.

Nath.ᵉˡ Wells
Leo: Jarvis } Managers

Safe
MS
082.2
L844h

Lottery tickets

SECRETARY'S-OFFICE, *Boston June 20ᵗʰ 1788*

THIS certifies, that Lot, Number *fourteen* Township Number *twenty seven East* Division, containing *one hundred & sixty* Acres, was drawn by the within Ticket, No. *2* and that *Messrs. Jonathan Hamlinton, Ivory Hovey and John Lord of Berwick in the County of York* *&c.* are Owners thereof, as appears by the Records in this Office.

160 Acres

John Avery jun.ᵈ Secretary.

Maine State Library

12 A rare photocopy of one of two remaining unsold lottery tickets in existence

surveys. William Bingham was a wealthy and patriotic gentleman from Philadelphia who contracted during the period 1786 to 1794 for three large tracts of public land, of approximately one million acres each, in the District of Maine.

General Henry Knox had contracted with Massachusetts in 1791 to buy one million acres on the upper Kennebec and fifty-two townships east of the Penobscot. He could not fulfill his obligations and signed the contracts over to Bingham. What thus became *Bingham's Kennebec Purchase* covered a total of about 1,128,000 acres with deductions from grants. It is commonly referred to as the million-acre purchase. The Kennebec River bisects the tract, and when townships were laid off or surveyed, they carried the designations and abbreviations of Bingham's Kennebec purchase west of the Kennebec River (B.K.P.W.K.R.) and Bingham's Kennebec Purchase east of the Kennebec River (B.K.P.E.K.R.). Examples are T. 2 R. 4 B.K.P.W.K.R. and T. 2 R. 5 B.K.P.E.K.R.

Directly tied in with Bingham's next million-acre purchase is the often referred to Grand Lottery. In 1786 surveyor Rufus Putnam had been authorized by Massachusetts to lay off fifty townships as lottery lands between the Penobscot and Passamaquoddy (St. Croix) rivers. Each township was to be subdivided into 160-acres lots. The land offered under the lottery scheme would, it was hoped, bring in half a million dollars. Accordingly 2,700 tickets were printed at 60 pounds each, payable in part by specie and in part by approved securities. At the conclusion of the sale only 437 tickets had been sold for $87,400, just 52 cents per acre, involving 165,280 acres.

Bingham's purchase of the area east of the Penobscot, between it and the St. Croix River, included the lands left over from the Grand Lottery. After the transaction it was found that the area was 48,024 acres short of the one million intended. In 1793 Park Holland was directed by the land agent of Massachusetts to lay out a two-mile and twenty-seven-rod-wide strip, forty miles long east and west (about the width of six townships) just north of the original main area. Similar strips of townships were also laid off east and south of the main tract. In the end the purchase covered 1,107,396 acres.

Thus townships involved in *Bingham's Penobscot Purchase* today carry the land survey designations and abbreviations of M.D. (middle division), N.D. (northern division), S.D. (southern division), and E.D. (eastern division). Examples of townships laid off can be identified as T. 34 M.D.B.P.P.; T. 4 N.D.B.P.P.; T. 9 S.D.P.P.; and T. 27 E.D.P.P.

Bingham took an option in 1793 on another million acres known as the *Back Tract*, north of Knox's original Penobscot tract. General Henry Jackson had contracted to buy it in 1791 and then later cancelled the purchase. The laying off of its townships was done by Park Holland and Jonathan Maynard in 1794. However, Bingham did not exercise this option.

The *Titcomb Survey* (*T.S.*) was made in 1794 by the Commonwealth of Massachusetts for the sale of Eastern Lands in public domain.

The Indians lost title to many large ungranted lands in public domain in the District. However, when Massachusetts, following the Revolutionary War, claimed ownership in fee simple to the vast areas of public lands in what is now the State of Maine, she recognized the cooperation of friendly Indians and worked out several treaties with them.

By order of the Massachusetts court in 1796 a tract known as the *Old Indian Purchase* (O.I.P.), consisting of a 30-mile strip on both sides of the Penobscot River, was to be laid off as part of a treaty with the Penobscot Indians. In 1797 surveyors Park Holland, Jonathan Maynard, and John Chamberlain did the actual surveying of this tract, consisting of 189,426 acres "more or less." In 1818 Massachusetts bought back this area, including the islands in the Penobscot River. The townships then laid off are now incorporated towns, but carry the original land survey designation of O.I.P.

Farther up the Penobscot River, land was also considered valuable for settlement. The following is quoted from Philip Coolidge's *History of the Maine Woods*:

> Lands up the river soon proved valuable for settlement, but the Indians, who had continued to be nearly destitute, misunderstood what rights they had sold or retained. Besides hunting and fishing, they had sold timber, and had attempted to make sales of lands to settlers. Accordingly, in 1818, by the "New Indian Purchase" the Indians gave up assumed rights on lands suitable for white settlement further up the river except for four townships, namely Mattawamkeag, Woodville and two townships somewhat west of the River still known as the "Indian Purchase." *

In 1833 Maine purchased these four townships from the Penobscot Indians for $50,000 and ultimately sold them in 1883 to private interests. This sum was set aside in the State Treasury. Interest from this principal is paid off to the Indians annually and continues to this day. The trust, which includes income from early sales of pine timber by the Penobscot Indians, as well as purchase of the four townships, amounted to $94,663.58, as of December 31, 1977.

The *North of Waldo Patent* (N.W.P.) land survey designation was

---

* Bangor: Furbush-Roberts Printing Co., Inc., p. 547.

the result of twenty-one townships surveyed between 1792 and 1842, and involves portions of lands known today as Penobscot, Piscataquis, Somerset and Waldo counties. The survey was ordered by the Commonwealth of Massachusetts as part of opening up the Eastern Lands to market. The names of Ephraim Ballard and Samuel Weston appear in the records as the early surveyors of these townships, followed later by Silas Holman, Alexander Greenwood, Rufus Gilmore, Samuel Redington, and other surveyors who did much of the lotting for settlement.

North of Bingham's Kennebec Purchase (N.B.K.P.) designates a continuation of surveys made north of the Bingham lands as part of its disposition of the wildlands of Maine. Actual field work appears to have taken place between 1811 and 1835.

North of Bingham's Penobscot Purchase (N.B.P.P.) designates another group of townships surveyed as part of the overall plan in the disposition of public lands by Maine and Massachusetts. The name of the surveyor Silas Holman appears frequently in the Land Office records for 1822.

The *Monument Line*: W.E.L.S. (west of the east line of the state) was run in the period 1825–1833 westerly from the point near North Amity where the east line of the state begins its due north course from the head of the St. Croix River.

The following description appears in the Land Office Records as the starting point of the Monument Line: ". . . a yellow birch tree marked & hooped with an iron hoop, starting at the head of the R. St. Croix near which stands a cedar post being about 10 inches square & twelve feet high above the surface of the ground which appears to have been placed there by the Commissioner under the Treaty of Ghent. . . ." Hence the name Monument Line. It is ninety miles long in a straight line to the east line of Seboomook (T. 4 N.B.K.P.) above Moosehead Lake, and covers fifteen townships.

This line was intended to be a base line for the location of townships in the public domain because of the contemplated division of these lands between Maine and Massachusetts. When the division agreement was reached in 1822, all townships north and some distance south of this line were laid off or surveyed west of the east line of the State, with each township assigned the land survey designation W.E.L.S.

The surveyors C. Norris, J. C. Norris, and Hiram Rockwood ran this line and also laid off a number of townships. The line is plainly indicated on U.S. Geological Survey maps.

West of Bingham's Kennebec Purchase (W.B.K.P) is, again, the designation for more of the laying out of tracts of land in public domain. Records show that in 1835 townships west and north of Bingham's Kennebec Purchase had been surveyed and divided between Maine and Massachusetts.*

Closely allied to the laying out of the great blocks of townships, and directly involving ownership and management of those forest lands which were to come under the supervision of the M.F.D., are the subjects of land titles and reserved public lots.

A substantial area of the land in the District today is held by owners who have a "common and undivided interest in the land." Such a system of land ownership is not unique to the State of Maine but is certainly unusual. This type of ownership is little understood by the public at large and is often confusing.

It should be understood that ownership "in common and undivided" does not constitute a title in and to some specific acre or acres within a township, but designates a fractional ownership together with some other fractional ownership of each and every acre or tree within that township. It is a situation that can be compared to that of a shareholder of a corporation. The Timber Resources of Maine Report of 1971 expresses this unique system very well:

> With this form of ownership, no division lines were drawn, and each of the owners held his personal undivided share of the total. Gains and losses from the ownership of the land also shared according to each owner's interest in the total. Thus, if some of the timber on the township were harvested, each owner would receive his proportionate share of the proceeds even if all the timber was harvested from one part of the township. Conversely, if a fire, insect outbreak, or other natural catastrophe struck the township, each owner would share in the loss according to his proportionate share.

Early deeds to individuals could be easily identified as a conveyance of a fractional interest such as a half, a fourth, a sixth or an eighth of a township. However, as families multiplied or individuals passed away, a complex system of fractional interests developed through inheritance, bequest, or devise.

Thus in the course of time these fractional ownerships were divided as a continuous process in varying proportions. A typical illustration of this process is explained by Louis S. Cook of the Great Northern Paper Company: "Let us take a certain deed of 1863 that conveyed a 31/52nd part of a one-eighth interest in common and undivided in a certain township. Following through the chain of title, we find that some present-day owners have a 332/33,280th holding in common and undivided in that township."

In some instances, these fractional ownerships are converted to a decimal that reaches six, seven, or eight places.

In spite of these fractional interests, the system works surprisingly well, although it makes for considerable bookkeeping for owners and for the state where each such fractional owner must be billed separately for taxes. Present owners show little inclination to alter this situation.

In all the discussion thus far on the disposition of public lands in what was the District of Maine and later became the State of Maine, nothing has been written about the "Public Reserved Lots." This subject, however, is an integral part of the story of deeds, grants, and sales of the public domain. The history of the public lots is well documented, but controversial issues have arisen in recent years that justify a brief background explanation of how they came into being and their purpose.

In 1784, one year after the establishment of the Massachusetts Land Office, an act was passed by Massachusetts requiring reservation of lands for the benefit of schools and established clergy, and providing that in the conveyance of each township there would be set aside two hundred acres for the first settled minister, two hundred and eighty for the use of a grammar school, and two hundred acres for the future disposition of the General Court. There is no record that any township was sold in Maine with this reservation.

In 1788, Massachusetts changed the conditions for conveyance of each township six miles square, requiring the reservation of "four lots of three hundred and twenty acres each [one] for the first settled minister, one for the use of the ministry, one for the use of schools, and one for the future appropriation of the General Trust." The latter became known as the "state lots." They were sold in fee simple and thereafter lost their identity except in the Land Office records. Thus, since all the so-called State Lots were disposed of at one time either singly or jointly by Maine and Massachusetts, the tabulation of reserved lands lists 960 acres in each township rather than 1280 acres. This reservation remained in effect until 1824 when Maine and Massachusetts jointly ratified a change to a thousand-acre contiguous public lot in each new six-mile square township.

When Maine became a state on March 15, 1820, the Articles of Separation became a part of the Maine Constitution. Article Seven provides the following: ". . . in all grants hereafter to be made by either state of unlocated land within the said District of Maine, the same reservation shall be made for the benefit of schools, and of the ministry, as has heretofore been used, in grants by this Commonwealth."

The Articles of Separation further provided division of ". . . all the public lands within the District between the respective states, in

Four Public Lots Reserved in a Township

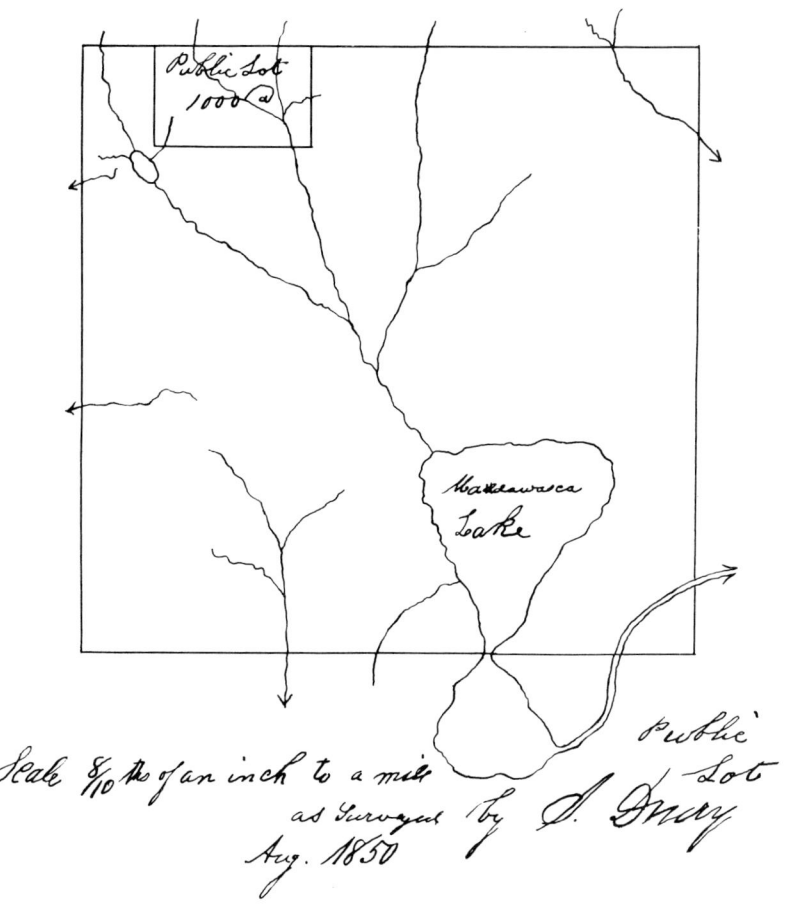

Map Showing 1,000-Acre public lot

KNOW ALL MEN BY THESE PRESENTS,

That I, *A. P. Morrill* Land Agent of the State of Maine, by virtue of authority vested in me by an act of the Legislature of this State, entitled "An Act in relation to lands reserved for public uses," approved August 28th, 1850, and in consideration of ———— *Two hundred fifty* ———— dollars to me paid by *S. P. & H. Strickland* of *Bangor* in the County of *Penobscot* the receipt whereof I hereby acknowledge, have granted, bargained and sold, and do by these presents bargain and sell unto the said *Stricklands* ~~his~~ heirs, executors, administrators and assigns, the right to cut and carry away the timber and grass from the reserved lots in Township *numbered five in the fourteenth range of Townships in the County of Piscataquis*

excepting and reserving, however, the grass growing upon any improvements made by any actual settler, said right to cut and carry away said timber and grass to continue until the said township or tract shall be incorporated, or organized for Plantation purposes, and no longer.

TO HAVE AND TO HOLD, the same as aforesaid to ~~him~~ *them* the said *Stricklands* ~~his~~ *their* heirs, executors, administrators and assigns.

IN WITNESS WHEREOF, I the said *A. P. Morrill* in my said capacity of Land Agent as aforesaid, have hereunto set my hand and seal, this *tenth* day of *April* in the year of our Lord, A. D. 185*2*.

SIGNED, SEALED AND DELIVERED
IN PRESENCE OF

J. R. Clark   *A. P. Morrill* seal

Penobscot ss. *April 17 1852* Then personally appeared *A. P. Morrill* Land Agent of Maine, and acknowledged the above instrument by him signed to be his free act and deed. Before me,

Isaac R. Clark   **Justice of the Peace.**

Example of a land agent's grant to a citizen of the right to cut and carry away timber and grass from unused reserved lots in a township

equal shares, situation and quality; they shall determine what lands shall be surveyed and divided, from time to time . . ."

In 1822, the above indicated division was accomplished, and in 1853 Massachusetts conveyed to Maine all its interests in lands in Maine, a total of 1,198,330 acres, at a price of $362,500, or a little more than thirty cents per acre. (See schedule of Maine Wildlands 1878, p. 8.)

The Schepps Report of the Attorney General's office (1972) gives an excellent and thorough historical perspective on public reserved lots. Since then there have been continuing questions and interpretations of existing statutes and suggestions for needed changes. Some have already occurred but are not updated here. Available material beyond the cut-off date of 1972 is considered not to be within the scope of this particular study. However, specific data and information can be obtained from the Bureau of Public Lands, Department of Conservation, Attorney General's office, and in legislative records.

By necessity this summary of the disposition and demarcation of Maine's forest lands has been brief. Those who may wish to pursue the sequence of events and major developments during that era in which a once vast public domain was converted to private and corporate ownership will find the story well documented. The Maine records are to be found in the files of private families, college and foundation libraries, and at the State Bureau of Public Lands, as well as in the Maine State Archives.

Of particular pertinence to this work are the records of the Maine Land Office, which run back to the 1820s. Up until recently these records have been under the custodianship of the Forestry Department. All the valuable books of maps; volumes of original field notes made during the running of exterior lines marking out townships and the subsequent lottings within these townships; field notes pertaining to the public reserved lands; deeds and grants were safely kept for years in a fireproof steel and cement vault within the Department. Recently, by act of Legislature, these records have been transferred to the custody of the Maine State Archives.

Services such as providing copies of maps, field notes, and plans are now carried out by the State Archives. The Land Office and Forestry Department's filing order and accompanying indexes are still retained by the Archives and provide information needed by surveyors and those working in title search. A most helpful brochure explanatory of the Land Office records is now available from the Maine State Archives.*

Of particular interest is the microfilming project conducted in

---

* See Appendix I.

1954–1955 by the Mormon Church in which certain Maine Land Office records were copied for the purpose of genealogical research. This project was part of a nation-wide endeavor. All the collected records are stored in the world's largest storage facility beneath a mountain in the Wasatch Range near Salt Lake City, Utah. The Forestry Department received a gift set of the microfilms taken of their records, and these are safely stored for posterity.

As brief as the preceding history has been, it will be helpful in the understanding of the present position of the State of Maine as well as of the conditions which brought the Maine Forestry District into being. It must be surprising to many that the M.F.D. is still largely composed, with the exception of a few municipalities, of a vast contiguous area of unorganized territory of a little over ten million acres.

It is not too commonly known either that 50 per cent of the annual timber harvest in Maine comes from the unorganized townships (most of which comprise the M.F.D.). There are many who do not understand why this annual cut is not larger, because the forest area is greater than in the organized towns (municipalities). The answer lies in the sustained yield management policies established and maintained by the landowners, as well as in the economic factors of accessibility of forests, roads, labor and transportation to the lumber and pulp-mill markets.

For many years (1920–1971) the State Tax Assessor by statute required returns from all timberland owners of the timber cut on their holdings in the unorganized territory for each fiscal year. The basic purpose was to help determine the revaluation of forest lands. Factors of growth increments and annual withdrawals of timber harvests were applied. Aerial surveys sampling a number of townships were also made periodically. A tabulation showing the annual cut by fiscal years for lumber, pulpwood, posts, ties and ship knees was kept up through 1971. This has now been discontinued as a result of the new Tree Growth Tax Law (Chapter 616, P.L. 1972 Maine). The annual cut will now be reported by calendar year to the Forestry Department.*

According to the State Tax Assessor's office, there are approximately eight thousand timberland owners, of which sixteen are large corporations, whose holdings lie within the M.F.D. There are still four large areas belonging to families descended from the "timber barons" of the past.

It must be emphasized that the timberlands within the District have for many years been under some form of forest management. The Seven Islands Land Company, which manages the holdings of the Pingree Heirs as well as several other family owned lands, is a good

---

* See Appendix I for complete tabulation of the last timber cut by species and volume for July 1, 1970 to June 30, 1971.

1973 State Valuation for all the Unorganized Territories $<u>130,661,603</u>
Prepared by Robert Meskers
State Bureau of Taxation

Areas <u>outside</u> the M.F.D. but <u>within</u> the Unorganized Territory

County	Location	Acreage	1973 State Valuation
Aroostook	Connor	23,952	652,683
Kennebec	Unity	6,255	100,178
Knox	All Islands	1,266	309,620
Lincoln	All Islands & Hibberts Gore	1,622	243,650
Oxford	Milton	8,794	119,107
Penobscot	Argyle	16,107	236,168
Hancock	<u>Islands</u>		
	Bald	5	2,900
	Bar	5	5,300
	Barred	15	12,840
	Birch	10	8,340
	Compass	1	2,500
	Eaton	16.7	5,070
	Fling	20	8,800
	Horsehead	8	4,200
	Inner Porcupine	9	4,400
	Little Marshall	2	3,600
	Outer Porcupine	6.5	4,400
	Pumpkin	3	5,830
	Scott	8	16,440
	Scrag	5	4,060
	Sheep	2	3,600
	Spectacle	3	2,700
Total Hancock Islands		119.2	94,980
Grand Totals		58,115.2	$1,756,386

Total Unorganized Territory outside M.F.D. $1,756,386
Total Unorganized Territory inside M.F.D. $128,905,217
                                             $130,661,603

example of such timberland management. Continuous records of the Pingree family are preserved at the Essex Museum in Salem, Massachusetts. They give records of forest management for a period of over a hundred years.

It is estimated that there are now more than one hundred industrial foresters employed within the state, with many of these working within the area protected by the M.F.D. The progress toward such professional management has been, in large, a recent occurrence. That such a concern is absolutely vital to the preservation of our forest is obvious. It is equally apparent that the interests of the various private owners and of the public at large across such a vast area could not have been served without an integrating program such as offered by the M.F.D.

UNUSUAL PHOTO OF LIGHTNING HIT ON A WHITE PINE STUB IN DENSE SPRUCE-FIR FOREST
Quick-action aerial water bombing held this fire in check until a ground crew arrived.
Photo taken at treetop level, T3R5 near Penobscot Lake, Somerset County, 1969

II

THE THREAT AND THE CONCERN

> *A fire devoureth before them: and behind them a flame burneth . . . behind them a desolate wilderness; yea and nothing shall escape them.*
>
> (*Joel* 2:3)

By the close of the nineteenth century the position of state land agent was an anachronism. In actuality the title lingered on until 1932, but only as an appendage to a new title that signaled a complete change in emphasis — from one of selling and disposing of that great public domain to one of its protection and preservation. In 1891, Cyrus Packard, who had served as land agent for ten years, became Maine's first forest commissioner.

Stewart Holbrook, writing in his *Yankee Logger,* makes the following statement: "The most striking thing about the forests of the northeastern United States is their persistence." There can be no doubt that the forest has been persistent, but that the grand treasure provided by the wilderness would persist under the ravages of fire, storm, and pestilence, combined with the increasing need of and utilization by man, had to be doubted.

A storm of hurricane force in September 1938 hit central New England, causing great havoc both in destruction of property and in wind-thrown timber. In Maine the area affected most was Oxford, Cumberland and York counties. Previous rains and the storm itself sufficiently loosened the soil to cause an estimated volume of over ninety million board feet of timber to go down, resulting in an extremely high forest fire hazard.

During 1939 and 1940 an excellent coordinated plan was worked out with private and public agencies in a gigantic merchantable timber salvage operation. Overall responsibility was assigned to the U.S. Forest Service, which functioned within the corporate structure of the Surplus Commodities Corporation. Organizational work was handled by a special setup known as the New England Timber Salvage Administration. Fire hazard reduction was carried out by the New England Emergency Project.

Private landowners received a fair market price for their down timber, with much of the manpower for salvage and hazard reduction provided by the W.P.A. and C.C.C. organizations.

One interesting aspect of this salvage project was legislative action authorizing the forest commissioner to use thirty-eight "great ponds" in the area for storage purposes. Existing sawmills were incapable of sawing logs to keep up with the rate at which trees were being cut and delivered to them.

Under the joint cooperative effort of private and public agencies, forty-eight million board feet were salvaged by the U.S. Forest Service and twenty-five million board feet by private owners. This is a remarkable 81 per cent salvage of the total volume of estimated down timber, and was largely due to the accessibility of most of the areas.

About 90 per cent of the timber salvaged was white pine and the remaining 10 per cent other softwoods. In addition, about 60,000 cords of pulpwood were salvaged. A total of $561,500 was paid for logs purchased. About three-fourths of the expenditures were for labor operations.

Completion of the salvage operation showed that much of the fire hazard had been reduced. In subsequent years, no fires of any consequence were reported in the areas logged over. One interesting point should be mentioned here. The fall of 1938 was an unusually good white pine seed year. As a result of this and the salvage logging operations, there was a good catch of pine seed, and regeneration began almost immediately.

In the M.F.D. there are records of timber blowdowns occurring in 1944, 1952, 1958 and 1964, but all were far less extensive in acreage and damage than was caused by the 1938 hurricane.

Early surveyors' field notes mention traveling through areas of heavy blowdowns, but specific information on these is lacking.

By the turn of the century the protection of the forests from its enemies and in particular the great destroyer, fire, was becoming a matter of public and private concern.

By the last decade of the nineteenth century, the forest fire problem in Maine had reached such proportions that the Sixty-Fifth Legislature in 1891 passed "An Act Creating a Forest Commission." Thus the first tentative steps were taken that would lead, eighteen years later, to the creation of the M.F.D. It was the threat of fire that first moved the Legislature and big timberland owners toward a joint concern.

Occurrence of forest fires in Maine was nothing new. In fact, from the first settlement and long before, fire had been the chief threat to stands of virgin timber. Undoubtedly, the very first forest fires were caused by lightning and by Indians, followed later by those set by settlers for the purpose of clearing land.

There has been much speculation concerning the susceptibility of certain forest trees over others to lightning hits. Several old proverbs advise: "Avoid the oak, flee the spruce, but seek the beech; beware the oak, it draws the stroke; avoid the ash, it courts the flash; creep under the thorn, 'twill save from harm." It does not follow, however, that the trees most likely to be ignited are the ones most responsible for spreading fire to the surrounding forest. In the records of the M.F.D. there are hundreds of known cases of spruce, fir, and pine trees as well as old snags set on fire by lightning that burned for days without spreading to other trees or on the ground. Some of these smoldering fires re-awaken after being dormant for a while. Most fires caused by lightning are due to the presence of dry duff, or humus, and litter at the base of the tree and to hollow dried-out dead snags and trees. In modern times, lookout towers and planes of the M.F.D. and private aircraft have spotted many "smokes" resulting from lightning hits, and hundreds of man hours have been spent by district wardens locating and extinguishing these fires. There can be no doubt of the destruction that must have been initiated by this natural cause in early times.

Indians are believed to have set the woods and fields afire for a number of purposes. It was a tactic employed to drive game into the open where a killing could be effected (unfortunately white men adopted this practice and it has not been completely eliminated today); to encourage the growth of wild edible berries; to clear land for agriculture; and to block enemy ambushers or raids.

The clearing of land by burning was also carried on extensively by early homesteaders. It should be remembered that there were few settlers in the vast interior and unorganized territory of Maine. The only other persons to traverse or occupy this country were explorers, surveyors, and loggers. A number of early fires can be attributed to these groups of people.

The amount of burnt land resulting from all causes was tremendous. The early records of the Land Office give frequent references to large burned-over areas. Thoreau in his explorations through Maine in 1846, 1853, and 1857 encountered such burns, and Austin Cary in his spruce studies of 1896, conducted on the Kennebec River, traveled

The fury of a forest fire

through burnt strips miles long and of considerable width. The "Report of Exploration West and Northeast of Katahdin . . ." by George H. Witherle, has the following items: "On September 16, 1899, large and numerous forest fires seen from the slopes of Katahdin. Immense smoke. A very large fire along the Wassataquoit, a large one near Millinocket Stream and Pamedumcook."

The very first of the early large forest fires of which records exist occurred in southern Maine in 1761–62. It originated in New Hampshire and swept across Maine to the sea in the vicinity of Falmouth and Portland. In 1796, a fire near Katahdin is estimated to have devoured 150,000 acres of prime spruce and fir forest. Interestingly, the date of this fire was ascertained through carbon-twelve analysis of a piece of charcoal found in an old peat bog.

By all accounts the fire of greatest historical interest and the one best remembered was the holocaust of October 7, 1825. It occurred in Piscataquis County and involved many townships, including Shirley, Elliottsville, Katahdin Iron Works, Long A Township, Kingsbury, Mayfield, Wellington, Harmony, Cambridge, and Ripley. Over 832,000 acres of forest and fields were burned over. The fire resulted from widely scattered fires, set by settlers to clear land, joining into one major conflagration fanned by a strong wind under extremely dry conditions.

One fact concerning the 1825 fire deserves mention. It is often mistakenly called the "Miramichi fire" in confusion with the *real* Miramichi fire, which occurred in the Province of New Brunswick, two hundred miles away from Piscataquis County but on the same day. The latter fire resulted in a heavy loss of timber along with the destruction of farm property. Two million acres were burned over, and a number of lives were lost.

Other large and destructive forest fires followed in the unorganized townships of Maine. One in 1837 is of special interest because it was set, reportedly, by a state land agent by the name of Chase. Having come across some fields of mown hay used as stock feed by poachers of timber, and thinking that the loss of the hay would slow down the poaching operations of the trespassers, Chase set fire to the haystacks. The result was a conflagration that burned over some 150,000 acres and nearly cost the agent his life.

In 1884, another fire occurred that swept over some 20,000 acres of timber and logging slash in the following townships: T.4 R.9, T.4 R.8, T.3 R.8, T.3 R.7 and T.4 R.7. This time the fire started a mile below the old "City Camp" in T.4 R.9, from a smudge set by fishermen in an attempt to drive away mosquitoes.

Other burns occurred in 1858, 1870, 1886, and 1899 in widely scattered places in the wilderness. The record of these is rather fragmentary and offers no specific information.

Then came the rash of fires in 1903. In the unorganized territory 200,232 acres burned, resulting in damages of approximately $761,583. A quick look at the statistics reveals that there were eight major fires that year, involving from eight to twenty-six thousand acres, raging at the same time. There were one hundred and thirty-six fires in all, of which four exceeded 20,000 acres in size. Forest Commissioner Edgar E. Ring in his report for that period wrote: "The forest fire record of 1903 will go down in history as one that has never been equaled, and it is hoped never will be repeated." (The 1947 fire disaster eclipsed this record, but most of the fires during this year took place within towns.)

The following quote, taken from an old record, illustrates the tremendous loss of natural resources occasioned by all these fires:

A little while ago Luther Rogers, the oldest lumberman in Patten, told me about a trip to the top of Katahdin which he made in 1856. He said he climbed Turner Mountain just this side of Katahdin and every tree in sight was old-growth spruce. East, west, north and south as far as he could see, the ground was covered by a heavy growth of spruce trees which were ten or twelve inches in diameter at thirty or thirty-six feet from the earth. The spruce is gone. The ground has been burned over. Only a thin growth of poplar is there now. The fire burned to the rock in most places so it will be another hundred years before it will be heavily wooded again.

Having provided an indication of the great threat of fire and the devastation which this enemy of the forests leaves in its smoldering wake, we shall turn again to the steps taken by the Maine Legislature and private landowners to establish a system of protection.

In January of 1891 the Legislature passed an order, "That the Judiciary Committee be instructed to inquire into the expediency of enacting some law for the better protection of the forests of Maine from fire and report by bill or otherwise." This request was concerned with all the forests of the state—in towns and cities as well as the wilderness.

At the same time, another order was proposed, "That the Committee on Taxation be directed to inquire into the expediency of providing by law that the State shall, in whole or in part, pay for the expenses of the fire departments of towns and cities in the State and also for the protection of wildlands of the State against forest fires." Although this piece of legislation did not pass, subsequent bills were enacted which did provide financial assistance.

In early March of 1891, the Committee on Judiciary responded to the legislative order by submitting a bill entitled "An Act to Create a Forest Commission and for the Protection of Forests." The enactment was not without the usual legislative process of tabling, reconsideration, and amendments, but it was finally signed into law on March 15, 1891, by Governor Edwin C. Burleigh. This new law provided for four specific areas of administration:

1. The State land agent was to become the forest commissioner.
2. Selectmen were made forest fire wardens and required to report fires.
3. County commissioners were to appoint forest fire wardens in the unorganized territory and to report fires.
4. The forest commissioner, with the approval of the superintendent of common schools and the president of the State College of Agriculture and the Mechanic Arts (now the University of Maine, Orono), was to awaken interest in forestry in

the public schools, academies and colleges with some degree of instructions.

As is so often the case, good intentions, even when embodied in laws enacted by the Legislature are ineffectual or even inoperative when funds to implement them are lacking. Such was the early situation faced by the first forest commissioner. The Law of 1891, which created the position, carried no authorization for funds with which the department could employ a force of fire wardens. In the case of the unorganized territory, the appointment of forest fire wardens was left with the county commissioners. By statute the primary responsibility of the forest commissioner was to compile forest fire statistics and conduct inquiries into the causes of, and means of preventing, fires—a most difficult task seeing that the commissioner had no men of his own in the field, and the reports of fires of two acres or more that county commissioners and selectmen were required under the statute to provide were very fragmentary and more than often never done.

A note from the Commissioner's Report of 1895 bespeaks the situation: "As the law now stands, the commissioner is powerless to act in cases of emergency — having no men at his command or means to employ them, excepting the fish and game wardens and they are so few and their beats so long, they cannot be depended upon unless the fire is in their immediate vicinity."

This arrangement was far from satisfactory. While the county commissioners were required by law to appoint fire wardens when they deemed it necessary, not to exceed ten in each county having unorganized territory, they often failed to comply or report their appointments to the forest commissioner. Where such appointments were made, wages were paid by the county. Fire fighters' pay came half from the county and half from the landowners.

An interesting tabulation has been prepared of the first appointments made in 1891 by the various county commissioners, along with the reported acres burned. It shows only twenty-nine wardens ap-

```
              1891 UNORGANIZED TERRITORY

     County          Acres Burned    Wardens Appointed
  Aroostook             18,662             10
  Franklin                 200             --*
  Hancock                  300              6
  Oxford                     0              2
  Penobscot                  5              3
  Piscataquis               16              6
  Somerset               2,400              2
  Washington             4,500             --*
                        26,083             29
```

*No notice to forest commissioner

pointed to cover an area of over 10,000,000 acres of unorganized territory.

A list of the wardens follows:

AROOSTOOK COUNTY — John Dunbar, No. 7, R. 8; Avon D. Weeks, Smyrna; George L. Byron, Linneus; E. R. McKay, Ashland; Haws, Ashland; Neal McClain, Ft. Francis; John A. Grant, Dyer Brook; John McAlwee, Presque Isle; Millard Filmore, Mapleton; A. B. Smart, Houlton.

HANCOCK COUNTY — Sumner W. Leighton, Cherryfield; Alfred Archer, Aurora; Nahum Jordan, Aurora; John R. Shuman, Great Pond; Joseph Clark, Ellsworth; George Watts, Beddington.

OXFORD COUNTY — Frank P. Thomas, Andover; Fred A. Flint, Wilson's Mills.

PENOBSCOT COUNTY — George F. Burleigh, Patten; George W. Fiske, Mattawamkeag; Frank L. Scammon, Lowell.

PISCATAQUIS COUNTY — Leonard Hilton, Chesuncook; George C. Luce, N. E. Carry; J. W. Ham, Day's Academy; Alphonso Bradeen, Lily Bay; Thomas C. Hamlin, Lake View; Charles H. Randall, Katahdin Iron Works.

SOMERSET COUNTY — David Butler, Flagstaff; H. Lincoln Colby, Jackman.

It was inevitable that certain weaknesses in the 1891 law would be corrected. In 1903, the Maine Legislature, acting on recommendations, put the first real teeth into the forest fire protection program in terms of money and authority. Two amendments to the original act of 1891 established an emergency fund of $10,000 for forest fire fighting and gave the forest commissioner the authority to appoint fire wardens in the unorganized territory.

This was most timely legislation, for 1903 proved to be a bad year, as has been indicated previously. The emergency fund was soon expended and even exceeded. Payment of bills had to be delayed until 1904, which fortunately was an extremely light year.

The emergency fund of $10,000 was later increased to $20,000, a figure that was to remain in effect until 1909, when the M.F.D. was created and all forest landowners were assessed a mill tax in proportion to their respective forest land valuations.

During the period of 1903–1908, the annual emergency appropriation was not a carrying account. All unexpended monies were returned to the State Treasurer and thus into the General Fund. Large fire bills had to be carried over into the next year, as in the case of 1903. Often these bills were scaled down by fifty per cent or more, and any cost above that paid by the state was borne by the landowners.

The forest commissioner had problems in adjusting claims by the landowners against the state on fire suppression bills. Often they were resolved by the state paying labor bills and the landowners the supply bills. Most landowners were willing to pay their proportionate part when there were insufficient state funds.

Under the amended law of 1903, the forest commissioner appointed one hundred and forty-one fire wardens. These men were paid two dollars per day for actual service. Fire fighters were paid *fifteen cents an hour*. It was difficult to recruit such men during busy times of year, especially when they could earn two to three dollars per day on the log drives and in other spring occupations.

The problems of the first forest commissioners were many, as illustrated by the following quotes from a letter written to me by Elizabeth Ring, the daughter of Commissioner Edgar Ring, on November 13, 1973:

> Recently you asked me what I remembered of my father's association with the Maine Forestry District Act of 1909 passed during his tenure as Maine's Forest Commissioner. Since I was very young and the specifics of the law were anything but within my comprehension, there isn't much that I can write except to recall vaguely and for me pleasantly some of the ways our family life was influenced by his ten years in the State House from 1900 to 1910.
>
> The first eight years of my life were marked by his activity in Augusta which took him away from home four days in the week. Our mother, after eight years of teaching in Pueblo, Colorado, had returned to Maine to marry my father, a childless widower of Orono whom she had known when earlier teaching there, and in no time at all to everyone's surprise presented him with three sons and a daughter. His life as Commissioner was exciting to her and his experiences were shared. Table talk on his return home every week was as likely to turn to the spruce budworm and the white pine blister as to the pork and beans of a lumber [camp], or what the news of the town was during his absence. It wouldn't be correct to say that the same cookies were in the cookie jar as when he left four days before, or that such an eventful four days with his associates in Augusta could fade completely as he re-entered the routine life of a family that in a sense was new to him. Neither was I always glad to see him, for I often took strong exception to being ejected in the middle of the night from sharing my mother's bed because the hired girl had said Mr. Ring had telephoned that he would be home on the midnight train.
>
> Forest fires were an ever present dread. As a practical lumberman operating for thirty years on the West Branch of the

Penobscot before 1900, and returning to timberlands after his tenure in the State House, that was the shadow that hung over the family occupation for three seasons of the year. From late spring when snow had left the ground early and dry winds with little rain had made the woodland a tinder box, to late October at the end of prolonged drouth, the danger of fire was ever there.

My earliest remembrance of hot sultry weather was more the wilted appearance of my father as he walked up Mill Street from the depot in Orono on a hot July day sweltering in a dark wool suit despite his Panama hat after a three hour ride in the grime of a train that stopped at every station. During the course of such a drouth the wall telephone was cranked frequently with the familiar drone of a toll call made to central and another crank when the call was completed. Early I was aware that fire wardens were men who put fires out. When the call went out to hire for the season, some who applied came to the house. One reluctant husband was brought by his wife who when my mother opened the door said, pointing to him, "He wants a job." Something different turned up when millionaire Jay Cooke, Jr., son of the Philadelphia financier, was given a commission as warden to fight fires at ten cents an hour. Cooke had a hunting lodge and extensive holdings in the upper Moosehead Lake region and a launch that plied easily between the shores of the Lake, and a telephone. Not that Cooke ever turned in a bill.

It was evident at an early age that some connection existed between fire wardens, land owners, and my father's absence from home with the need to make speeches which he rehearsed in the big room upstairs. Sometimes I sat on the top stair and wondered what the fun was talking out loud in an empty room. Then, too, his absence during the week drew the family more closely together on Sunday evenings knowing that the early morning train would take him away again. Stories of hunting and fishing were told and often repeated and interspersed with incidents that had happened over the week. Mr. Lannigan of Waterville had taken him up country in his Stephens-Duryea. There was chaos in the countryside, as chickens scattered in all directions and horses reared and snorted. One horse tethered to a post supporting the roof of a piazza, at the noise of the engine broke loose and pulled with him post, piazza, washing and all, to the fury of the owner. On another occasion returning from Calais on a Washington County train, the engine struck three deer killing them. It was open season and the game warden aboard took a ribbing as having brought down more than his quota. Frequently Father talked of the men with whom he had had dealings and if they were incompetent or stupid, as an illustration his eye would rove around the family circle until it fell on me. At the time I expect I enjoyed the attention.

Compensation for these weekly absences were the trips we took to Augusta, much more exciting for the train ride than the sight of the State House of which we soon tired. The Commissioner's right hand man was Mr. Charles Curtis of Brewer. Being childless, he and Mrs. Curtis made [much] of us and were frequently dinner guests as were the instructors of Forestry at the University of Maine when the chair was established by Governor Cobb in 1903. Both encumbents during these years, Professor S. N. Spring of the Yale School of Forestry, and Gordon Tower who came from the Pacific Northwest, were often at the house with their wives. Mr. Gordon was a very tall man whose face seemed to me to be as long as the rest of him. Both did much to improve the forestry service in the state, as did Austin Cary of Bowdoin, and later of Harvard, who wrote scientific articles for the Commissioner's biennial report.

But the high spot of these years came when my father, with Cary and Ex-Governor Hill attended President Theodore Roosevelt's Governor's Conference on conservation held in the East Room of the White House in May 1908. This buoyed the family a bit and among the several hundred delegates who attended, it was fun on his return to pick him out in the sea of look-alike faces. What transpired in praise of what Maine had done for the protection of its forests was something that was passed around in the family and enjoyed.

Politics never entered into anything I ever sensed about my father's job. But the day was coming. In the early evening of the September election of 1910, when my mother and I were taking her favorite walk through the bridges, men were loitering on street corners and reclining on lawns as we passed. She explained that it was election night and the Democrats who wanted the repeal of the Maine Prohibitory law would probably win, in which case another man would be appointed Forest Commissioner instead of my father. A year later the shingle of the E. E. Ring Land Company went up in the Kirstein Building overlooking the Kenduskeag River in Bangor. At the age of sixty-three my father entered the third period of his life in a field of activity associated with the woodlands of Maine which he dearly loved and about which through experience and study he had learned a great deal.

With the appointment of fire wardens by the forest commissioner there began a systematic plan to create a number of forest fire districts based on river systems or other convenient boundaries. Later on it became an established policy to appoint fire wardens by watershed districts, such as the Saint John Waters, Aroostook River, Kennebec Waters, etc.

From this beginning, a pattern of a well-defined series of forest fire protection districts began to form. The skeleton force of seasonal wardens, however, was not adequate and many large landowners continued to hire their own patrolmen and watchmen to supplement the state's effort.

A quote from the Forest Commissioner's Report of 1903–1904 illustrates increasing action taken by the landowners themselves in contrast to their previous attitudes in which forest fires were considered just one of the risks to investment that had to be expected and tolerated:

> In the spring of 1903, the landowners on the waters of St. John and Allagash united in employing four men to do patrol duty on these waters. These men were put on the work the first of May and continued on said work till October 1. One man was located at St. Pamphile, P.O., near the boundary, and covered the region around Big Black River; this region is infested with Canadian settlers near the boundary who make a business of making sugar on the American side each spring, and need close attention. The other men patrolled the St. John River and its tributaries in canoes, following the drives down the river and the sportsmen and driving crews up. They put out all camp fires which were left burning. These men were appointed fire wardens by the forest commissioner, and so they had authority to call for help in case of a fire getting beyond their control. This is the first time anything of this kind has been tried and it has proved very satisfactory, inasmuch as no fires of any extent occurred during the season. The expense of four men was about $160 per month, borne entirely by the landowners, who paid in proportion to their interests in the several townships covered by the patrol.

To strengthen the forest fire protection program, the landowners continued their cooperation by paying for the construction of three lookout towers in 1905 — Attean Mountain, Mount Bigelow, and Squaw Mountain — at a cost of $750 each. Salaries of the watchmen were paid by the state. During the period of 1905–1908, six additional towers were built from landowner funds.

By 1908 the foundation for a better system of forest protection had been laid through the action of the Maine Legislature and through the increasing concern of the landowners. Perhaps the basic change in the attitude of all concerned is best illustrated by the suggestion a fire warden made in the year 1904: "Impress upon the people the benefit of keeping the forests green." The warden's suggestion was seminal. Years later "Keep Maine green" and "Keep America green" were to become the battle slogans of a program to preserve the great forest resource.

It is well to point out that forest conservation began, not with the Federal Government, but with the programs initiated within individual states, and Maine was one of the pioneers in this area. (Maine's dedicated Forest Commissioner Edgar E. Ring was one of the men called to Washington, D.C., in 1909 to help shape a national forest conservation policy.)

In 1908, the stage in Maine was set for further progress. It was a year to precipitate increased action. There were one hundred and twenty-six fires, with 98,691 acres consumed and damages amounting to $361,796. The situation called for the first "woods ban" or closure by a proclamation of the governor, and amidst this anxiety the M.F.D. took form.

FIRST AND LAST FOREST COMMISSIONERS

Edgar E. Ring　　　　　　　　　　　Fred E. Holt
Forest Commissioner 1909　　　　　Forest Commissioner 1973

First and last forest commissioners to serve during the sixty-three
year period of the Maine Forestry District

III

CREATION OF THE M.F.D.

*The ideals cherished in the souls of men
enter into the character of their actions.**

The enactment of the legislation leading to the formation of the Maine Forestry District is one of the most interesting pieces of forest fire control legislation in the history of this country. For the first time there began in Maine an era of well-organized fire protection for the unorganized territory of the state. The M.F.D. became, as one landowner aptly stated, "the preventative maintenance and control of forest lands against fire — a sort of defensive maintenance." The system has been widely acclaimed as a model for administering and financing protection against fires and other threats endangering large areas of private forest lands.

Despite the importance of the Maine Forestry District in the annals of forest protection as a prime example of the cooperation of private and public interests, it is surprising how few, aside from those directly concerned, know its full significance. This is especially astonishing in the case of some legislators who still think that M.F.D. means "Maine Forestry Department." In light of this, a proper history of the M.F.D.'s creation seems in order.

Between 1891 and 1908, certain significant events occurred that eventually led to the legislation of 1909 and the M.F.D. Chief among these were the two disastrous fire years of 1903 and 1908, which resulted in heavy losses of timber and high suppression costs. Under such conditions the attending problems became vividly apparent. There was no uniformity in the hiring of fire wardens, the token "emergency fund" of $10,000 for paying patrolmen and watchmen was obviously inadequate. Moreover, landowners had increasingly large

---
* Alfred North Whitehead, Philosopher.

expenses for the protection of their own lands through patrols, lookout towers, and cost-sharing with the state in the expenditures incurred in extinguishing fires. It was widely felt that such expenditures could be better equalized and used within a unifying organization. Finally, an aroused public opinion demanded that positive action be taken to protect the forest resources from fire. The increase in public concern is evidenced in the following quote from the Forest Commissioner's Report of 1905–06, "It is not, however, altogether the owners of timberlands who shudder at the knowledge of a forest fire, but more and more our people are realizing what general destruction of our forests by fires would mean."

In this sobering situation, the conviction materialized that the interests of the state and her people would be better served by some form of stringent legislative action that would create more adequate protection of the vast areas of the forest land. The M.F.D. was the result of that conviction.

So far as we know, there was no one individual whose efforts resulted in the M.F.D., but the combined thinking of many landowners led to a recommended plan for better forest fire protection and the creation of a District concept. While records are meager, there is a fair amount of certainty that Forest Commissioner Edgar E. Ring had the support of such individuals as Fred A. Gilbert, Woods Department manager of the Great Northern Paper Company; Blaine S. Viles, consulting forester, landowner, and later forest commissioner; Hosea B. Buck, manager for the Pingree Heirs; and J. P. Bass, a Bangor publisher.

Irving G. Stetson, in a letter written to me, gives an interesting personal glimpse of Fred A. Gilbert, along with a list of other notable men who were probably influential during the M.F.D.'s formative period:

> I am sure that Fred A. Gilbert, the first Manager of the Woods Department of the Great Northern Paper Company, was one of the leading proponents of the act, if not the one primarily responsible for it, as I recall that both my father and uncle, who were very close to Fred A. Gilbert and the latter's father, Thomas Gilbert, gave me such an impression. My father and uncle "staked" Thomas and Fred A. Gilbert on their logging operations for six to seven years starting in 1891, and both of them were always very friendly to my father and uncle; and Fred A. was always telling Father that it was due to the confidence which Father and Uncle Edward had in him (Fred) that he had made a success in life.
>
> I myself got to know Fred A. Gilbert (I always include the

"A" when referring to him, as I have another very good friend, Fred Gilbert, of Greenville, whom I have known for 40 years) quite well commencing in about 1920, when the first of my three sons was 6 years old, as, although Fred A. Gilbert was considerably older than I, he had married very late and his children were of the same ages as ours, went to the same school and played with our boys. Fred A. Gilbert was, in my opinion, quite an outstanding man, although one without much formal schooling, and was the best possible man whom the Great Northern could have chosen for that job in the early years of the company's history. Some people considered him cold-blooded, but, while he would not stand for anybody "pulling a fast one on him" and never forgot it, he would "lean over backwards" to repay anybody who did a good job for him or rendered him a favor. He never forgot a man who was above-board with him and did a good job, and would go out of his way to show his appreciation. . . .

Outside of Fred A. Gilbert, I cannot give you the names of any others who I am sure might have "initiated" the Forestry District Act, but there may have been some in the following list:

Edgar E. Ring, Forest Commissioner
Roy L. Marston, Manager for Coburn Heirs
Blaine S. Viles
Charles W. Mullen
John Cassidy
James W. Cassidy
Dr. Charles E. Adams
Fred A. Powers
Eugene Hersey
Josea B. Buck, Manager for the Pingree Heirs
Edward Blake

J. P. Bass
B. B. Thatcher
George B. Dunn
Philo A. Strickland
Frederic H. Strickland
Charles V. Lord
William Engel
Henry Prentiss
Samuel R. Prentiss
Frederick A. Appleton
Edward R. Godfrey
J. P. Webber
Charles P. Webber

The names above include the large landowners and operators of that day.

Elizabeth Ring adds several names in one of her letters to me:

At the time of the fiftieth anniversary of the law [the enactment of the M.F.D. legislation], I gathered some material on my father's service to the state as Forest Commissioner. . . . Father's secretary was a man named Charles Curtis, of Brewer, who has long since passed away. As has also a lawyer in Winthrop named

Carleton who was associated with Father on the Fish and Game Commission. John and Charles Oak were also names that I could associate with my father's activity during this period. . . .

It is quite possible that some of the first chief wardens appointed in 1909 by the forest commissioner could have participated in the discussions on the proposed legislation.

Chief Wardens Appointed in 1909

AROOSTOOK WATERS
George B. Dunn, Houlton
S. C. Cummings, Haynesville
William Sewall, Island Falls
Harry E. Hasey, St. Francis
Cony A. Pooler, Old Town
Fred C. Knowlen, Guerette
Ora Gilpatrick, Houlton
J. B. Bartlett, Ashland
H. B. Buck, Bangor
Wm. H. Hinckley, Bangor
Eugene H. Decker, Bangor

HANCOCK COUNTY
H. T. Silsby, Aurora

KENNEBEC WATERS
W. M. Shaw, Greenville
Louis Oakes, Greenville Jct.
E. P. Viles, Skowhegan
W. J. Lanigan, Waterville
Frank E. Haines, Deadwater
Forrest H. Colby, Bingham
F. H. Sterling, Caratunk
Blaine S. Viles, Augusta
Peter Herbst, The Forks

OXFORD AND FRANKLIN COUNTIES
Silas F. Peaslee, Upton
C. C. Murphy, Rangeley

PENOBSCOT SYSTEMS
John Appleton, Bangor
Eugene H. Smith, Norcross
M. L. Woodman, LaGrange
J. A. Obley, Mattawamkeag
S. C. Cummings, Haynesville
Fred A. Gilbert, Bangor
John W. Hinch, Danforth
N. C. Ayer, Bangor
Chas. W. Bowers, Mattagamon
J. L. Chapman, Milo
S. H. Boardman, Guilford
A. R. Billings, Brownville
E. O. Grant, Patten

WASHINGTON COUNTY
Victor M. Smith, Northfield
Thos. O. Hill, Codyville
Alfred K. Ames, Machias

Although it is difficult to state just which of the above men was most influential, the general thinking that went into the drafting of the original M.F.D. bill is clear.

In 1909, a figure of at least $50,000 was recommended by Forest Commissioner Edgar Ring as necessary for an annual budget for forest protection. Many discussions were held by a select group of landowners as to how best to go about raising this amount of money. Admittedly, it would be unjust to the other interests of the state to ask for an increased appropriation from the General Fund. To avoid any charge of unfairness, it was decided that the burden should be placed on all the landowners in the unorganized territory in the form of a self-imposed tax prorated on an acreage basis. A series of consultations was held between the landowners, their legal counsel, and members of the Legislature. All reacted favorably toward this approach, including the Committee on Taxation, which consisted of:

Wheeler of Cumberland
Macomber of Kennebec   in the Senate
Mullen of Penobscot

Wing of Kingfield *
Additon of Leeds
Trickey of Corinna
Pattangall of Waterville   in the House
True of Portland
Richardson of Presque Isle
Colby of Bingham **

* Herbert S. Wing
** Forrest H. Colby, forest commissioner 1917–18; 1919–20

The answer was a draft proposal calling for a special forest fire tax to be levied on all landowners within the unorganized territory. Such a tax would remove any objection of unfairness in regard to payment and would relieve the state of any fire control costs on these lands. The landowners would undertake the whole burden of protection from forest fires — each in proportion to acreage owned — and all would act unanimously. This would eliminate anyone getting a "free ride." This proposal was submitted to the Maine Legislature.

On February 22, 1909, Senator Carl Millikan of Island Falls (later governor 1917–1920) introduced a bill entitled "An Act Creating a Maine Forestry District and Providing for Protection Against Forest Fires Therein." This bill was referred to the Committee on Taxation. Later it was ordered to be printed under the joint rules.

At this point the Committee on Taxation raised the question of constitutionality and the bill was referred to the Judiciary Committee, the members of which were:

Hastings of Oxford
Looney of Cumberland   in the Senate
Baxter of Cumberland **

* Percival P. Baxter, donor of Baxter State Park and former governor

Davis of Yarmouth
Peters of Ellsworth
Hersey of Houlton
Montgomery of Camden } in the House
Andrews of Augusta
Burleigh of Augusta
Wing of Auburn

Some historians link the epic and unique ruling of the Maine Supreme Court on the regulation of cutting in 1908 as contributory to the enactment of the M.F.D. This decision was based on the fact that it was within the rights and powers of the state to regulate, on behalf of the public, the manner in which private forest lands were *managed* and *protected*. The Court rendered its decision at the request of the State Senate, but it never became law.

The following is an often quoted statement:

We think it a settled principle, growing out of the nature of a well ordered society, that every holder of property, however absolute and unqualified may be his title, holds it under the implied liability that his use of it shall be so regulated that it shall not be injurious to the equal enjoyment of others having an equal right to the enjoyment of their property, *nor injurious to the rights of the community.*

It is to be noted that the constitutionality of the Maine Forestry District was again to be challenged in an interesting court case, "Inhabitants of Sandy River Plantation vs Weston Lewis and Josiah S. Maxcey of Gardiner and owners of property in Sandy River Plantation." In 1912 a rescript drawn by Associate Justice George F. Haley, of Biddeford, ruled that the District was constitutional.*

Finally, after further study, the bill as amended was passed by the Senate on March thirty-first (24 *yes* to 0 *no*) and by the House on the same date (108 *yes* to 21 *no*). After a lengthy debate in the House, Amendment "A," which carried an emergency clause for assessing a tax rate of one and one half mills for the year 1909, was also passed by the House (103 *yes* to 0 *no*). The bill was signed into law by Governor Bert M. Fernald on April 1, 1909, as Chapter 139, Section 1-15, Public Law, Maine, 1909.

Basically the new law provided for:
(1) A small mill rate with mandatory annual assessments upon all landowners based on valuation owned by each.
(2) An assured income to establish and maintain an orderly forest fire protection organization.

---

* See Forest Commissioner's Report, 1912, pages 13–18, at Maine State Library.

MAINE FORESTRY DISTRICT 1971-72
Total acreage by classification

399	townships	8,342,751.28 acres
	public lots	317,431.67 acres
	miscellaneous	244,788.50 acres[1]
43	plantations[3]	1,096,117.00 acres
12	Incorporated towns[4]	322,979.00 acres[2]
		10,324,067.45 acres*
		161,800.00 acres**
	Total Acreage M.F.D.	10,485,867.45 acres

*Figures taken from Maine State Valuation 1973 and Planimeter Survey, Maine State Planning Board 1935
**Adjusted figure for M.F.D. of increased forest area from 1971 Maine Timber Resources Report based upon improved forest inventory, data processing techniques, aerial photos and land use changes.

[1]Includes: tax delinquent and tax exemption, Baxter State Park, lands unconveyed and State owned. (Reference: Maine Municipal Association)
[2]Data from planimeter survey by State Planning Board 1935 and still used as a reference. (Reference: Maine Register 1974-75)
[3],[4]Plantations and Incorporated Towns came into the M.F.D. at different times by legislative action. (Reference: Maine State Valuation 1973)

TABLE OF FOREST AREA IN MAINE
(Updated Changes)

	Early Figures (Prior to 1959) Acres	1959 Figures Acres	Latest Figures (1971-72) Acres
Organized Towns	6,429,783	7,103,933	7,264,733
M.F.D.	10,262,455	10,322,067	10,483,867
Totals	16,692,238	17,426,000*	17,748,600**

*Initial Maine Timber Resources Report – Organized Towns and M.F.D. computed
**Second Maine Timber Resources Report – Computed to include 3 per cent increased forest area of 323,600 acres in 1971 Report

```
1971 - 17,748,600 Acres
1959 - 17,426,000 Acres
         322,600 Acres
```

Acreage increase in 1971-72 is the result of improved forest inventory, data processing techniques, aerial photos and land use changes. Work sheets are available to show how figures were computed at the Department of Conservation.
See Appendix II for breakdown of territory by counties, and Table of Municipalities Joining the M.F.D.

TABLE OF MUNICIPALITIES
(ORGANIZED TOWNS AND PLANTATIONS)
JOINING MAINE FORESTRY DISTRICT AND GROSS ACRES OF FORESTS

	Year Joined	Gross Acres		Year Joined	Gross Acres
NORTHERN REGION			WESTERN REGION		
Aroostook County:			Franklin County:		
Allagash Plt.	1909	86,400	Coplin Plt.	1909	21,632
E. Plt.	1909	13,939	Dallas Plt.	1909	24,512
Garfield Plt.	1909	23,386	Rangeley Plt.	1909	25,490
Glenwood Plt.	1909	23,712	Sandy River Plt.	1909	22,445
Hammond Plt.	1909	25,343			94,079
Macwahoc Plt.	1949	15,840			
Nashville Plt.	1909	21,108	Oxford County:		
Oxbow Plt.	1909	22,048	Lincoln Plt.	1909	22,778
Reed Plt.	1949	34,963	Magalloway Plt.	1909	33,385
Wallagrass Plt.	1949	25,261			56,163
Westmanland Plt.	1909	22,285			
Winterville Plt.	1909	21,606	Piscataquis County:		
Subtotal-Northern		335,891	Bowerbank	1949	20,390
			Medford	1949	25,574
EASTERN REGION			Barnard Plt.	1909	16,256
Hancock County:			Elliotsville Plt.	1909	31,891
Osborn Plt.	1965	22,918	Kingsbury Plt.	1909	27,418
No. 33 Plt. (Great Pond Plt.)	1909	23,962	Lakeview Plt.	1909	25,299
		46,880			146,828
Penobscot County:			Somerset County:		
Medway	1949	23,386	Moose River	1909	25,798
Drew Plt.	1909	22,272	Moscow	1949	29,101
Grand Falls Plt.	1909	21,753	Brighton Plt.	1965	25,792
Lakeville Plt.	1909	33,625	Caratunk Plt.	1949	34,093
Seboeis Plt.	1909	23,386	Dennistown Plt.	1909	25,593
Webster Plt.	1909	22,246	Highland Plt.	1909	27,558
		146,668	Pleasant Ridge Plt.	1909	16,141
			The Forks Plt.	1909	26,285
Washington County:			West Forks Plt.	1909	32,781
Beddington	1949	20,410			243,142
Centerville	1949	26,208	Subtotal-Western		540,212
Cooper	1949	20,160			
Crawford	1949	22,131	GRAND TOTAL		1,385,088
Deblois	1949	23,405			
Northfield	1949	28,013			
Topsfield	1949	30,611			
Wesley	1949	31,885			
Baring Plt.	1965	8,845			
Codyville Plt.	1909	32,083			
Grand Lake Stream Plt.	1909	26,771			
No. 14 Plt.	1909	19,053			
No. 21 Plt.	1909	25,862			
		315,437			
Subtotal-Eastern		508,985			

(3) Full responsibility and authority vested with the forest commissioner.
(4) Establishment of forest fire protection districts based upon waterways and other natural boundaries.
(5) A free hand in employing supervisory and field personnel, in the construction of telephone lines, lookout towers, and warden living quarters, and in purchasing of equipment.

The newly formed M.F.D. included in its responsibility the unorganized townships, some plantations, the coastal and inland islands, and by subsequent legislation some contiguous municipalities. To be added in the years that followed would be Baxter State Park, Allagash Wilderness Waterway, Rangeley State Park, Indian Township, all of which are located within M.F.D. boundaries, and the Public Reserved Lots. In 1909 the area of the District was approximately 9,500,000 acres. Later this figure was revised to a little over 10,000,000 acres as a result of more accurate ground and aerial mapping surveys, an increase which also included an extension of coverage due to legislative action.

It is interesting to point out that in 1909 the initial tax rate of one and a half mills provided $63,945.44. With this fund, the forest commissioner started his M.F.D. forest fire organization. Needless to say, these funds were quickly put to work with nearly half the amount used for increasing the number of patrolmen. The balance went for lookout towers, equipment, vehicles, etc. The forest fire tax of eight and a half mills in 1972 made available $1,331,161.69 based upon a valuation of $157,736,629. Between 1909 and 1972 there were twelve mill tax fluctuations. The growth in revenue can be attributed to increased valuation within the District, years of high suppression costs and rising costs in salaries, added personnel, equipment, insurance, retirement payments, and capital improvements. Today it is big business to administer and operate the forest fire protection program that began with the creation of the M.F.D. in 1909.*

---

* See Chapter V and Appendix II for further tables relating to forest fire taxes.

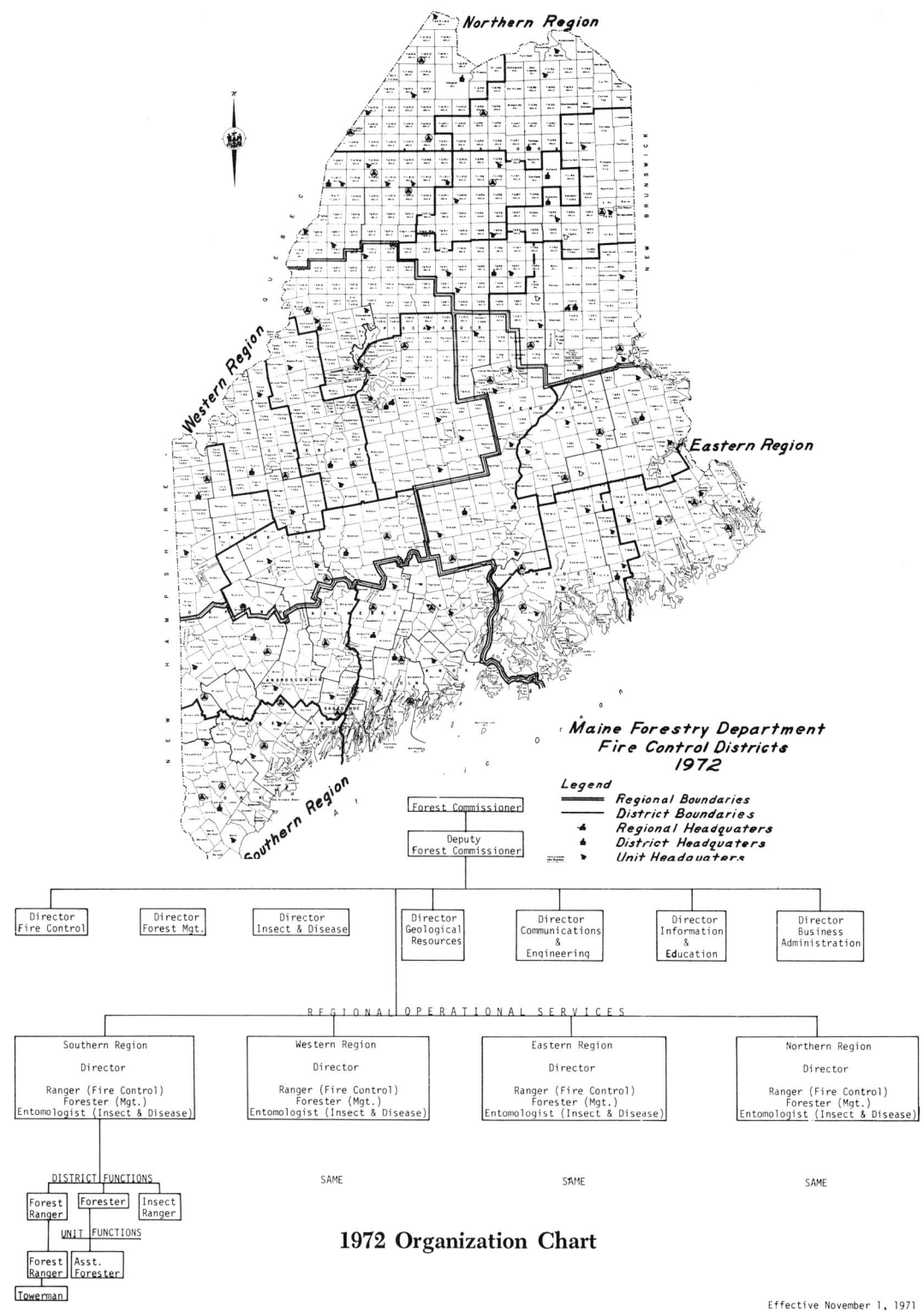

1972 Organization Chart

Effective November 1, 1971

IV

ORGANIZATION AND WARDEN SERVICE

If you plan for one year—plant rice
If you plan for ten years—plant trees
If you plan for a hundred years—train men.
Old Chinese Proverb

The basic intent of the Maine Forestry District as instituted under Maine law was to establish and maintain an effective administrative structure within the unorganized territory of the state strictly for forest fire protection. With the creation of the M.F.D., the forest commissioner had, for the first time, the necessary tools of authority and funds for setting up such a protective system. It was the start of an organization that was destined to grow and expand in the years ahead as the task of protecting over ten million acres of contiguous forest was met with increasing efficiency.

The mammoth proportions of this task can be readily appreciated from the enormity of the geographic area to be covered by both supervisory personnel and field forces. With virtually no roads penetrating the vast forest areas, it was natural to establish protection areas on the basis of watersheds and according to river systems and other convenient boundaries. Travel, in those days, was principally by canoe, making the delineation of subdivisions in relation to the network of lakes, rivers, and streams ideal.

Initially the M.F.D. was divided into six broad areas; Aroostook Waters, Hancock County, Kennebec Waters, Oxford-Franklin counties, Penobscot Waters, and Washington County. As time went on, these names changed, the boundaries of these large tracts became more refined, and their interiors were subdivided.

A good illustration of such further subdivision and refinement would be the big, general area first designated as "Aroostook Waters."

MAINE FOREST FIRE CONTROL PLAN - 1951
CENTRAL HEADQUARTERS
(Augusta)

Forest Commissioner - Deputy Forest Commissioner
Departmental Business Manager - 2 Dispatcher-Draftsmen - Radio Engineer and Assistant
Radio Dispatcher - 4 Clerical Staff - 1 Pilot

Maine Forestry District -- 10,262,455 A.   Organized Towns
                                                          6,429,783 A.

Northern Division 2,241,348 A.	Central Division 2,532,467 A.	Western Division 3,525,714 A.	Eastern Division 1,962,926 A.	Supervisor
Supervisor	Supervisor Asst. Supervisor	Supervisor	Supervisor	
6 Chief Wardens	6 Chief Wardens	7 Chief Wardens	6 Chief Wardens	6 District Wardens
18 Patrolmen	14 Patrolmen	22 Patrolmen	12 Patrolmen	1 Warden-Mechanic
10 Watchmen	23 Watchmen	25 Watchmen	16 Watchmen	24 Seasonal Wardens
6 Tel. Operators	5 Tel. Operators	7 Tel. Operators	2 Tel. Operators	30 Watchmen
33 Deputy Wardens	86 Deputy Wardens	122 Deputy Wardens	103 Deputy Wardens	446 Town Wardens
				1,338 Deputy Town Wardens

25 State Employed Full Time - 240 State Employed Fire Season - 2735 on Call

---

MAINE FOREST FIRE CONTROL PLAN - 1952
CENTRAL HEADQUARTERS
(Augusta)

Forest Commissioner - Deputy Forest Commissioner
Business Manager - 2 Dispatcher-Draftsmen - Radio Engineer and 2 Assistant
Radio Dispatchers - 5 Clerical Staff - 1 Pilot

Maine Forestry District -- 10,262,455A.   Organized Towns
                                                          6,429,783 A.

Northern Division 4,773,815 A.	Western Division 3,525,714 A.	Eastern Division 1,962,926 A.	Supervisor
Supervisor Asst. Supervisor	Supervisor	Supervisor	
12 Chief Wardens	7 Chief Wardens	6 Chief Wardens	7 District Wardens
32 Patrolmen	22 Patrolmen	12 Patrolmen	1 Warden-Mechanic
33 Watchmen	25 Watchmen	16 Watchmen	24 Seasonal Wardens
11 Tel. Operators	7 Tel. Operators	2 Tel. Operators	30 Watchmen
1 Pilot			
119 Deputy Wardens	122 Deputy Wardens	103 Deputy Wardens	445 Town Wardens
			1,335 Deputy Town Wardens

25 State Employed Full Time - 240 State Employed Fire Season - 2,735 on Call

The name of this area was later changed to Saint John Waters and subdivided into the Madawaska, Fish River, Allagash, Seven Islands, Upper St. John, Aroostook Waters, and Number Nine sub-districts. A similar pattern followed for the other big regional areas in the M.F.D.

It was Samuel T. Dana, forest commissioner from 1921 to 1923, who recognized the need for bridging the gap in coordination between the Augusta office and the regional chief wardens in the field. He recommended dividing the M.F.D. into four regional divisions, grouping the various protection areas which had been established under chief wardens, but still preserving the natural waterway boundaries of lakes, rivers, and streams, and other convenient boundaries. Neil Violette, who succeeded Dana, implemented his recommendation in 1925, forming the first organized and functional forest fire protection structure in the M.F.D., with a total personnel of 447.

The four divisions were labeled Eastern, Northern, Central, and Western. Each was handled by a district chief who later assumed the title of supervisor and who worked out of the Augusta office until the establishment of permanent field headquarters.

The position of supervisor was recognized as an important part of the chain of command. Oddly enough, the title does not appear in the statute, but it was the product and natural outgrowth of the M.F.D.'s expansion program. The supervisors have been, for the most part, technically trained men who have had experience in administrative planning, organization, execution of work plans, and supervision of personnel. Over the years the size of the supervisors' divisional areas has varied considerably, due largely to better modes of transportation and changes in personnel.

The four divisional areas of the M.F.D. are a good example of the organizational reform that increased the efficiency of administrative management between the Augusta office and the field force. The reorganization meant that the forest commissioner would be dealing with only a few top supervisory people rather than many, thus giving him more time to carry out the other duties of his office.

The four-division system continued up until 1952 when there was again a geographical regrouping that resulted in a reduction of the divisions to three—Northern, Western, and Eastern. This continued until November of 1971 when an entirely new reorganization plan went into effect that, while returning to a four-division system, entailed a far greater scope of coordination.

The following statement made by me as forest commissioner at the time indicates the importance and scope of the changes being made:

The new plan establishes for the first time an organizational structure which functions at a State staff level of directors serving in an advisory and support capacity. Then there is the line organization, where the action is, of four regional directors adminstering the basic programs of fire, disease, and small woodland management programs. Most significant is the fact that now the Maine Forestry District and Organized Towns become one entity. Programs will be administered in terms of protecting the total forest resources of nearly 18,000,000 acres and not under divided programs.

Within each region or division, the fire control organization of both the M.F.D. and the Organized Towns was combined with the other major activities of the Forestry Department. This interdepartmental development, in particular those aspects regarding forest fire control, was approved by the governor, select members of the Legislature (there was no general legislative action), and Forest Department personnel. There were the usual growing pains of adjustment, but the new structure proved its worth. Most importantly, its main thrust became one of protecting the *total* forest resources of the state under one unified organization.

At the apex of the M.F.D. organizational structure was the position of the forest commissioner, whose particular and vital role demands special attention. In common with many heads of state departments, he was appointed by the governor, an appointment which until recently has been subject to the approval of the Governor's Council. Under the statute (Title 12, sec. 501, R.S.), the appointee was to be a trained forester or a person of skill and experience in the care and preservation of forest lands. The term of office was four years.

The position of forest commissioner had some unique differences when compared to other state department heads. First, salary payment was derived from both the General Fund and the Maine Forestry District tax. Second, the commissioner was responsible to the governor only, with no intervening board, commission or advisory committee. And, third, the nature of the position meant that the forest commissioner serve on numerous time-consuming allied or related committees.

Some of my most pleasant experiences were my meetings with the governor. Through reviews or progress reports made at least once every two months, and through frequent meetings, the governor was able to gain a close insight into the functions as well as the problems of one of his most important departments.

This direct communication has now been dissolved through the creation of the new Department of Conservation (Chapter 460, P.L. 1973). The forest commissioner acts as director of the Bureau of Forestry, and as such is responsible to the commissioner of conservation.

It is doubtful if any other department head was required by virtue of office to serve on as many allied or related committees. At one time there were eleven such committees on which the representation of the forest commissioner was required by law or appointment. A few are listed here: the Maine Mining Bureau, Northeastern Forest Fire Protection Commission, Baxter State Park Authority, the Conservation Foundation, State Park and Recreation Commission, Eastern States Exposition Board of Trustees, the State Pesticides Board, the Indian Township Advisory Council, and the Land Use and Regulation Commission. Alternates at times filled in for the forest commissioner, whose own duties were more than often a full-time occupation.

The various positions forming the organizational structure under the direction of the forest commissioner can best be viewed by chronological periods, thus showing not only the chain of command but also the constant process of reorganization over the years made necessary by both technical and administrative progress.

As the forest fire organization of the M.F.D. grew and expanded, there were several periods when a series of different titles were used for certain positions in the fire warden service. Some carried several titles for the same position; some remained the same; others changed, while still others were dropped. An attempt has been made here to research through the records and to trace the nomenclature of titles once used as well as those now current for the periods 1891–1908, 1909–1924, 1925–1970, and 1971 to the present.

I. 1891–1908:

The blanket term "fire warden" appears for the first time in the Legislative Act of 1891 (Chapter 100, section 4. P.L. 1891) when the county commissioners were authorized to appoint at their discretion "such fire wardens" as were deemed necessary for the unorganized territory. The primary function of such appointed wardens was one of patrolling the waterways of lakes, rivers, and streams.

In 1903 the Legislature amended the act of 1891, removing the power of appointing fire wardens from the county commissioners and placing it with the forest commissioner. At this time the title of "patrolman" also came into use, arising from the wording of the act that gave the commissioner the vested power "to appoint forest fire wardens to patrol the forests." In 1905 the title of "watchman" came into being with the introduction of lookout towers.

II. 1909–1924:

In 1909, the creation of the M.F.D. provided for a much broader range of appointments by the forest commissioner to include chief

SCHEDULES OF NOMECLATURE CHANGES OF M.F.D. WARDEN SERVICE - 1925-1972

1925

Augusta	Forest Commissioner Asst. Forest Commissioner Clerical Staff
Section or Division	Supervisor Chief Warden Patrolman Watchman Deputies on Call
	Deputies - Railroad

About 1950s

Augusta	Forest Commissioner Dept. Forest Commissioner Radio Engineer Radio Dispatcher Pilot Clerical Staff
Division	Supervisor Asst. Supervisor
District	Chief Warden Asst. Chief Warden Patrolman Watchman Warden Mechanic Tel. - Danger Sta. Op. Deputies - on call
General	Deputies - Railroad General Deputies Honorary Chief Wardens

1966-1971

Augusta	Forest Commissioner Dep. Forest Commissioner Asst. Div. Forest Ranger (Fire) Business Manager Radio Supervisor Radio Dispatcher Property Accountant Clerical Staff
Division	Division Ranger Asst. Division Ranger Pilot Radio Technician
District	District Ranger Asst. District Ranger Forest Ranger Watchman Ranger Mechanic Aircraft Mechanic Tel.-Radio-Danger Sta. Op. Deputies - on call
General	Deputies - Railroad General Deputies Honorary Chief Wardens

1972-73

Augusta	Forest Commissioner Dep. Forest Commissioner Asst. Dir., Fire Control Dir., Business Management Dir., Radio Communications Radio Dispatcher Property Accountant Clerical Staff
Region	Regional Director Asst. Regional Director Regional Ranger (Fire) Ranger Pilot Radio Technician Ranger Mechanic Aircraft Mechanic Warehouseman Secretary
District	District Ranger (Fire) Ranger Watchman Tel.-Radio-Danger Sta. Op. Deputies - on call
General	Deputies - Railroad General Deputies Honorary Chief Wardens

forest fire wardens as well as deputy forest fire wardens, which included patrolmen, watchmen, and general deputies.

Heading the new setup from Augusta was the forest commissioner and his assistant, who later carried the title of "deputy forest commissioner," both these titles continuing until October of 1973.

In the field, a definite pattern of organization now began to form with the establishment of forest fire control districts, each under the charge of a chief warden. At one time there were thirty-two such districts. Interestingly, this number had been reduced to fifteen by 1971.

Early in this new structure the present title "chief warden" was known as "head" or "chief warden-at-large." Later a change had to be made to distinguish between those who took active leadership and those appointed under this title who were purely advisory to the forest commissioner (prominent landowners or agents of large paper companies). They became known as "honorary chief wardens" or "general deputies," and their appointments continue to this day. Those who were active on the M.F.D. payrolls and directly responsible to the commissioner were known by the more permanent title of "chief warden."

Chief wardens appointed by the forest commissioner up until 1925 were directly responsible to and received instructions only from him. In each district the chief warden was in full charge of all personnel patrolmen, lookout watchmen, and other deputy wardens, who worked only when called upon during periods of greater fire danger. The duties of the chief wardens were clearly spelled out in the law, and ever since the first appointments in 1909 have remained basically the same. Paramout to everything else, they were and are held accountable for the forest fire control in their respective districts.

From the top echelon of forest commissioner and his deputy, the field force within each warden district bore the title of chief warden, patrolman, watchman and deputies on call.

Around the year 1911, the title of "chief warden-in-charge of railroad patrol" appears in the records of the M.F.D. Appointments were made by the forest commissioner of those eight to ten railroad officials who were responsible for the right-of-way through which the tracks of their company ran. Under each chief warden-in-charge of railroad patrol were appointed deputy railroad patrolmen who were usually section foremen, each responsible for certain sections of railroad tracks. Hundreds of chief and deputy patrolmen were appointed annually by the commissioner. In 1967 this arrangement was changed, with patrols put on by railroad officials without certificates of appointment from the Augusta office.

Mr. V. J. Welch, manager of operations for the B & A Railroad Company, wrote to me as follows on January 26, 1967:

> I have your letter of January 23rd concerning the appointment of our track maintenance personnel as general deputy wardens and, frankly, I feel just as you do.
>
> We will, of course, cooperate with your people and the State of Maine Forest Service in every way possible. I feel that our cooperation will be as effective without the general appointment of our forces as deputy wardens, and I am sure that this will eliminate quite a bit of paper work on the part of both our organizations.
>
> I, therefore, recommend that we go through this coming season and see if there are any situations developing where this decision might not be the best and if there are we can reassess the situation another year. However, as stated above, I can see no reason why our cooperation will not be as effective and meaningful as in the past.

III. 1925–1970:

In 1925 a significant addition was made to strengthen the forest fire organization with the adoption of the position of "division supervisor." As has already been mentioned, the division supervisor served as liaison between the Augusta office and the district chief wardens in the field. Originally this title was known as "inspector" or "agent." For several years a separate title of "inspector" was applied to a few men appointed by the forest commissioner on temporary or special assignments to check and inspect lookout towers and the condition in general within each of the chief warden districts. These were summer jobs filled by one or two faculty members from the Department of Forestry at the University of Maine. Such positions were discontinued when the division supervisors moved into field headquarters.

Another position which appeared for a brief period of time was that of "lineman," a position carried during the peak years of heavy woods telephone maintenance. These men were specialists assigned a month at a time to each chief warden district to correct problems or to improve the telephone network system.

With the further increase of the duties and responsibilities of both supervisors and chief wardens new positions were inevitable. Into the organizational structure came the position of telephone-radio fire danger station operators, assistant supervisors, assistant chief wardens, aircraft pilots, warden mechanics, aircraft mechanics, radio technicians, and clerical help.

Growth at the Augusta level was also accompanied by new positions and titles—assistant division forest fire ranger, business manager, radio engineer, radio dispatcher, property accountant, and again, additional clerical staff.

The chart on the following page, showing the organizational structure in 1925, serves to illustrates the complexity that had developed over the sixteen years since the formation of the M.F.D. in 1909.

Final changes occurred in 1966 and 1969, in the establishment of a numerical range of positions within the fire organization in order to comply with the State Personnel Department. Special titles not in the forest ranger category remained the same. A comparison of titles appears below:

1966 Title Changes

Forest Watchman	Forest Watchman
Forest Patrolman	Forest Ranger I
Forest Warden II	Forest Ranger III
Asst. Supervisor	Asst. Supervisor
Forest Warden IV	Forest Ranger V

1969 Title Changes

Forest Watchman	Forest Ranger I
Forest Ranger I	Forest Ranger II
	Forest Ranger III (new)
Forest Ranger III	Forest Ranger IV
Asst. Supervisor	Forest Ranger V
Forest Ranger V	Forest Ranger VI

(*Note:* This forest ranger numerical arrangement was retained when the Department went over to a state-wide regional setup in 1971.)

IV. 1971 to the present (1973):

On November 1, 1971, a complete change was made, affecting both the M.F.D. and the services of fire wardens in Organized Towns. This change marked the end of each of the above organizations as separate entities and combined the two into one unified forest fire ranger system for the total forest area of nearly eighteen million acres.

The changes from 1925 to 1972 were all part of an administrative reorganization within the Forestry Department. It is to be expected that the legislative act creating a new State Department of Conservation (effective October of 1973) will bring about still more changes in positions and titles, but such transitions are outside the scope of this history of the M.F.D.

Organization of Maine Forestry Districts

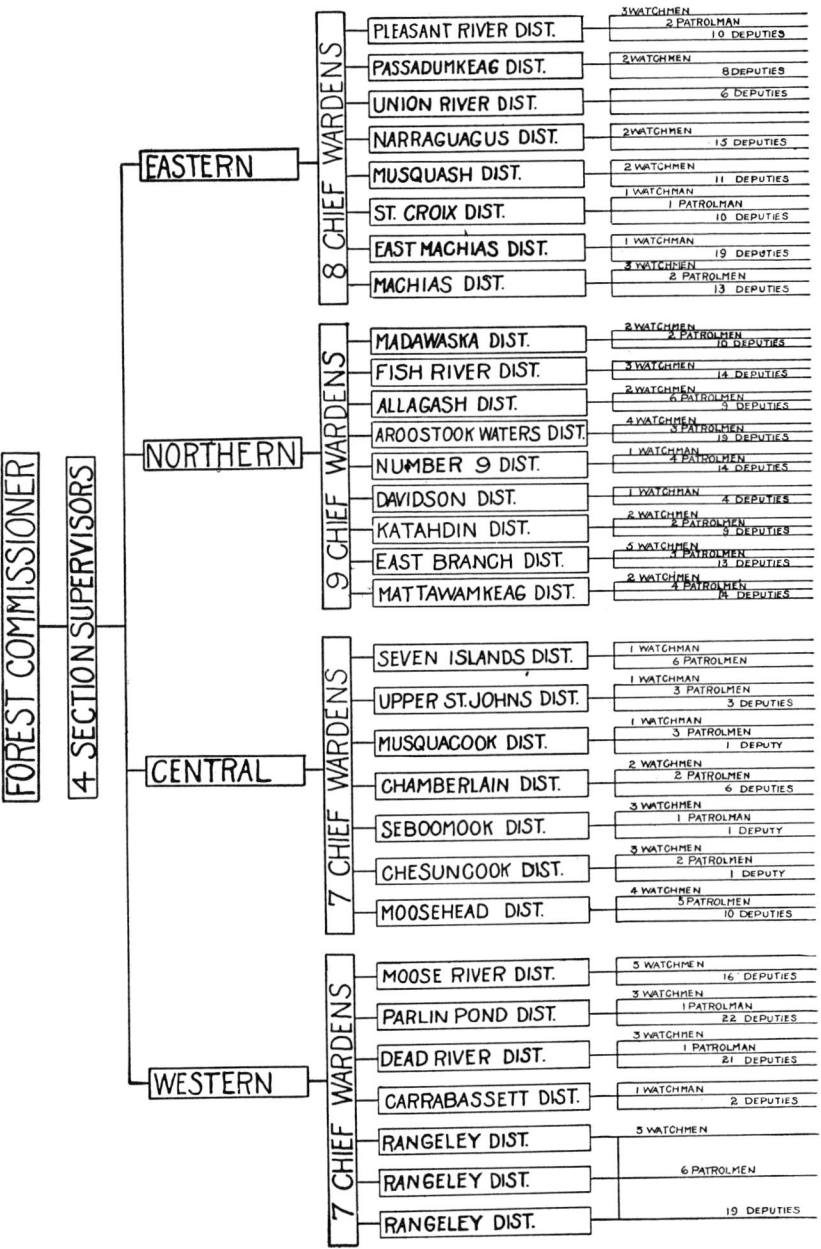

FIRST ORGANIZATION OF MAINE FOREST DISTRICT
Total personnel in 1925 was 447

As has already been pointed out, early patrol work in the District was primarily along the vast network of waterways. Those were the days of the long log and pulpwood drives down the rivers and of sportsmen who hunted and fished along the rivers and streams. These two groups were the main risk factors in the concern over forest fires.

Then came the era of increased cutting operations and of lumber camps created to meet the demand for more wood by the pulp and paper industry. At the same time, truck roads were built to reach remote areas, which until this time had been generally inaccessible, thus opening up the wildlands to the tourist and sportsman. It is to the credit of the landowners that they permitted the public to use these roads, but their use also increased the threat of forest fire. (Increased use of these roads, now approximately 3,000 miles, by the public often necessitated the imposing of restrictions when hazard from logging operations or danger of forest fires demanded such action.)

To meet this "opening up" of the unorganized territory, the Forestry District found it necessary to employ more personnel: patrolmen, watchmen, telephone-radio-danger station operators, and others to counter the increased risk of fire. The network of roads, while increasing the danger of fires set by careless people, also provided better access for M.F.D. wardens, extending their capabilities for prevention and suppression of fires. As new roads were extended into the woods so grew the warden patrol force. The timberland owners, recognizing this situation, provided added funds through increased fire taxes to meet the budget requests.

MAINE FOREST SERVICE FIRE PERSONNEL SUMMARY - 1961

Augusta	Maine Forestry District	Organized Towns
Forest Commissioner	3 - Division Forest Rangers	Division Forest Ranger
Deputy Forest Commissioner	2 - Ass't Div. Rangers	6 - Dist. Forest Rangers
Business Manager	18 - Dist. Forest Rangers	23 - Forest Rangers
3 - Forest Draftsmen	3 - Pilots	1 - Ranger Mechanic
3 - Radio Technicians	1 - Radio Technician	29 - Watchmen
1 - Radio Dispatcher	69 - Forest Rangers	19 - Danger Sta. Op.
7 - Clerical Staff	2 - Ranger Mechanics	445 - Town Wardens (On Call)
2 - Supervisors	53 - Watchmen	1,335 - Dep. Town Wardens (On Call)
16	24 - Danger Sta. Op.	1,858
	315 - Deputies on Call	
	490	

FORESTRY DEPARTMENT
FOREST FIRE CONTROL DIVISION FIELD PERSONNEL - 1971

Augusta	Organized Towns
Forest Commissioner	1 - Division Forest Ranger
Deputy Forest Commissioner	1 - Assistant Division Ranger
Business Manager	6 - District Forest Rangers
1 - Division Forest Ranger	23 - Forest Rangers
1 - Assistant Division Forest Ranger	1 - Ranger Mechanic
2 - Supervisors	25 - Watchmen
1 - Radio Supervisor	445 - Town Wardens on Call
3 - Radio Technicians	1,335 - Dep. Town Wardens on Call
1 - Radio Dispatcher	1,837
1 - Property Accountant	
10	

Home of the Forestry Department, sixth floor State Office Building, Augusta

V

FINANCING AND ACCOUNTING

> *Yet the unique pattern of fire control cost-sharing developed in Maine served as the model for other forest economy states all the way across the continent.**

One of the most important facets in the history of the M.F.D., as in all large operations, is that of financing and accounting. In the case of the M.F.D., it is an unusual story, a sixty-three years' record of a "special dedicated revenue account" being utilized to effect a forest fire protection system without any assistance from the state's General Fund. The primary source of revenue behind the M.F.D. and its operations was a self-imposed forest fire mill tax, rated on a dollar valuation, upon all the landowners whose forest holdings lay within the Maine Forestry District. Supplementing this revenue from the mill tax were federal grant-in-aid allotments under the Weeks Law of 1911 and the Clarke-McNary Act of 1924.

What is written here carries special significance by virtue of the fact that since the enactment of the Maine Tree Growth Tax Law (Chapter 616, section 1608, P.L. 1972) the M.F.D. has ceased to be a separate entity financed by the special tax for forest fire protection purposes; and all future funds for this purpose in unorganized territory will have to come from state General Fund appropriations. This is the net effect, but the eight and one half mill tax is still specified, and it goes through the General Fund appropriations process.

Early accounts and recent actions taken in the overall financial picture of the District are well documented. From such records two

---

* Ralph Widner—*Forest and Forestry in the American States.*

most noteworthy points become evident to the researcher. First, there is a common thread throughout the sixty-three year period—the emphasis on *cost-consciousness* and an awareness of the necessity of *accountability* in regards to funds raised for fire protection. Secondly, for the last two decades, at least, an excellent spirit of cooperation existed between the forest commissioner and the Maine Forestry District's advisory committee. Especially helpful was the financial subcommittee with Arthur Stedman of Scott Paper Company as its chairman.

The salient factor of cost-consciousness can best be illustrated by the following quote from instructions to chief wardens given by Forest Commissioner Neil L. Violette in 1932:

> However, I feel that we can carry on our usual work and still have a little balance, providing we do our bit in trying to use strict economy for the balance of the year. The permanent force of the office, which includes the supervisors and myself, will do its part by contributing two weeks salary, which means a saving large enough to pay the entire expense of one district for one month. Your bit should consist in holding down your expenses to the minimum and getting along with supplies and equipment that you have on hand. In other words, outside of gasoline for your cars and boats, do not spend one cent, unless it is absolutely necessary.

The following two letters from Commissioner Mace to John E. Mitchell, Chief Warden of Patten, Maine, reflect the same concern over curtailing expenditures. The first one was dated June 3, 1915:

> I have your May bill and note that you have used your *horse* 38½ days which I presume is to June 1st. After careful consideration I have made an allowance of one dollar per day to wardens using their own team, which I trust will be satisfactory. I have added $38.50 to your bill.

The second letter was dated July 22, 1915:

> Your letter of the 14th at hand. I have not promised a single man under you straight time and if they actually believe they were to be allowed this then they must have been misinformed. This includes Craney and Cote.
>
> I do not wish any man under you to charge for time during periods of wet weather when it is not necessary for them to work and I shall look to you to see that they do not. You are on the ground and know the conditions and should be able to judge whether they should be allowed for certain days or not and I will

stand behind you in whatever you do in this respect and I trust that you will use your best judgment in dealing with the same.

From my experience in patrol work I know that there are men that will try in every way to keep from receiving word to quit work. I have known instances where they would cut telephone line so that they might get in a few more days. The case of John Coote brings this to mind so be governed accordingly when you approve his bills.

As regards the Eastern and K.P. board bill the regular allowance made by this Department for boarding men is 50 cents per day and in no case on a fire has over 25 cents per meal been allowed.

A postscript was added in ink:

If you have work for Craney use him as he is a long way from home, and is a good woodsman.

The above quote and letters bear testimony to the constant effort of the forest commissioner and his staff to cut costs whenever weather and forest conditions allowed. The practice of laying off patrols during periods of heavy rains and wet forest conditions occurred during a time when recruitment of labor was not too difficult and the factors of fringe benefits and full-time employment were of no great concern. One can't help but reflect that by today's standards of employment, the policy of earlier days would never be tolerated. However, the practice of closely scrutinizing all expenditures early became a fixed policy in the M.F.D. and continued into more prosperous times.

No matter how carefully the funds of the M.F.D. were administered, the fluctuating costs of fire protection had to be faced and their increase over the years accepted. At the base of the financial structure was the forestry district tax, which in 1909 brought in $63,945.44 as a result of the one and one half mill per dollar valuation of woodlands. Sixty-three years later (1972), the tax of eight and a half mills resulted in a revenue of $1,331,161.69 based upon a valuation of $156,607,266. The full spectrum of the expanded services and their cost between the first year of M.F.D. operation and 1972 is exhibited in the following tables.

MAINE FORESTRY DISTRICT
Expenditures
Summary for 1909

```
Patrolling Account............................$23,090.05
Cost of Lookout Stations and Telephone lines..  7,380.15
Watchmen on Lookout Stations..................  4,893.54
Expense Extinguishing Fires................... 10,944.80
Chief Fire Wardens............................  6,510.99
Deputy Fire Wardens...........................  1,524.75
Tools and Supplies............................  2,163.25
Other Expenses................................  2,232.32
                                              $58,739.85
```

MAINE FORESTRY DISTRICT
EXPENDITURES
Fiscal Year Ending June 30, 1972

	Budget F.Y. 1972	Total Expenditures June 30, 1972
Fire Suppression Costs	$ 60,000.	$ 29,807.71
Personal Services	860,995.	802,720.62
Special Services	2,400.	3,477.42
Travel Expenses	17,500.	21,346.77
Operation of Vehicles	71,500.	74,324.34
Operation of Planes	35,000.	35,651.31
Utility Service	19,600.	20,713.68
Rents	48,750.	50,644.97
Repairs	33,950.	23,739.94
Insurance	35,982.	31,313.57
General Operating Expense	12,100.	12,141.81
Food	500.	318.39
Fuel	8,500.	6,286.35
Office Supplies	4,050.	3,768.84
Clothing	9,000.	10,016.04
Supplies & Small Tools	22,550.	30,295.75
Grants to Cities & Towns		2,520.77
Purchase of land		2,500.00
Buildings & Improvements	32,000.	6,470.64
Equipment	138,510.	191,199.51
Structures & Improvements	6,000.	3,224.34
Retirement Contributions	101,225.	101,546.56
TOTAL EXPENDITURES	$1,520,112.	$1,464,029.33

MAINE FORESTRY DISTRICT
FINANCIAL STATEMENT
Fiscal Year Ending June 30, 1972

	Operations	Spruce Budworm Control	TOTALS
Comptroller's Balance July 1, 1971	$1,743,929.42	$ 41,144.35	$1,785,043.77
RECEIPTS:			
Maine Forestry District Tax	1,331,161.69	366,612.61	1,697,774.30
Federal Grants	361,621.60	2,013.00	363,634.60
Rent of Buildings	3,543.09		3,543.09
Witness Fees	3,047.00		3,047.00
Misc. Services and Fees	10,326.76		10,326.76
Miscellaneous Sales	249.39		249.39
Miscellaneous Receipts	14,401.00		14,401.00
Serv. Fees Chgd. Other Depts.	3,283.71		3,283.71
Fire Protection-Baxter Park	12,000.00		12,000.00
Fire Protection-Passamaquoddy	2,040.00		2,040.00
Fire Protection-State Parks	1,238.64		1,238.64
Sale of Buildings	1,050.00		1,050.00
Sale of Equipment	5,683.47		5,683.47
Sale of Autos	729.00		729.00
Sale of Clothing	9.78		9.78
Sale of Supplies	105.80		105.80
Settlement of Fire Losses	771.59		771.59
Misc. Rents & Leases	4,438.00		4,438.00
Other Settlements	932.80		932.80
Adjustment of Balance Forward	292.97		292.97
Contribution from General Fund		300,000.00	300,000.00
TOTAL RECEIPTS	$1,756,926.29	$668,625.61	$2,425,551.90
TOTAL AVAILABLE	$3,500,855.71	$709,739.96	$4,210,595.67
Less Expenditures to 6/30/72	1,464,029.33	319,898.26	1,783,927.59
BALANCE 6/30/72	$2,036,826.38	$389,841.70	$2,426,668.08

The levying of the fire tax rate was applicable throughout the unorganized territory of the M.F.D. and also in some areas outside the District, such as a few unorganized towns and the forty-nine coastal islands.* In the case of latter, the revenue was credited to the Organized Towns account for administration. The following table serves to show the classification of those areas taxed and the computation by which the totals were derived.

*See Appendix V for breakdown of State Valuation for all unorganized territories.

MAINE FORESTRY DISTRICT 1971 TAX COMPUTATIONS
(Rate at 8½ Mills for Operations)

	Valuation	Tax Revenues
Unorganized towns	$125,532,962[1]	$1,067,030.18
Buildings on leased property	5,338,030	45,373.25
Public lots	3,545,870	30,139.89
Subtotal	$134,416,862	$1,142,543.32
Organized towns	23,319,767[2]	198,218.02
Total - Operations	$157,736,629	$1,340,761.34

[1] Excludes income of $10,480.63 with a valuation of $1,233,015 for unorganized territory and islands outside of the Maine Forestry District.
[2] Computed valuation.

MAINE FORESTRY DISTRICT TAX ASSESSMENTS*
Forest Fire Tax Assessment

Year	Authorization	Mill Rate	Amount	Valuation (Computed)
1909	Chap. 193, P.L. 1909	1 1/2	$ 63,945.44	$ 42,630,293
1919	Chap. 105, P.L. 1919	1 3/4	112,773.87	64,442,211
1921	Chap. 5, P.L. 1921	2 1/4	157,043.56	69,797,137
1949	Chap. 103, P.L. 1949	8	484,319.84	60,539,980
1951	Chap. 90, P.L. 1951	5 1/2	347,840.44	63,243,716
1953	Chap. 2, P.L. 1953	9 1/2	669,170.68	70,439,018
1954		5 1/2	387,428.23	70,441,496
1955	Chap. 13, P.L. 1955	4 3/4	463,095.70	97,493,831
1958		4 3/4	472,090.50	39,387,475
	Chap. 424, P.L. 1957	1 1/2**	118,361.31	
1960		4 3/4	498,790.11	105,008,444
	Chap. 376, P.L. 1959	3/4**	65,555.62	
1963		4 3/4	524,052.66	110,326,875
	Chap. 5, P.L. 1963	2 1/4**	207,386.85	
1965	Chap. 102, P.L. 1965	5 1/4	593,548.14	113,056,789
1967	Chap. 29, P.L. 1967	9	1,021,188.02	113,465,335
	Chap. 101, P.L. 1967	1/2**	48,200.16	
1968	Chap. 29, P.L. 1967	8	907,819.75	113,477,469
1969	Chap. 190, P.L. 1969	8 1/2	1,075,045.51	126,475,942
1970		8 1/2	1,083,510.69	127,471,846
	Chap. 533, P.L. 1969	1 **	106,104.23	
1971		8 1/2	1,340,761.34	157,736,629
1972		8 1/2	1,331,161.69	156,607,266
	Chap. 617, P.L. 1971	2 3/4**	366,612.61	

*See Appendix I for table showing annual M.F.D. assessments 1909-1972 as taken from the biennial reports.
**Special assessments for Spruce Budworm Spray Projects

For further clarification, the M.F.D. tax was based upon a rate fixed by the Legislature and valuation as appraised by the Board of Equalization. In Organized Towns within the M.F.D. territories, beginning in 1958, personal property and buildings were excluded from the valuation base.

Between 1909 and 1972, there were twelve forest fire mill tax rate changes. As we have seen, the rate in 1909 was set at one and a half mills, and this figure remained in effect for the next ten years. The rate increased in 1919 to one and three quarter mills and to two and a fourth in 1921, where it remained for twenty-eight years. Since 1949, the mill rate has changed frequently from a low at four and three quarters to the last mill rate of eight and one half in 1969.

The total assessments levied against all landowners within the M.F.D. between 1909 and 1972 reached the impressive cumulative figure of $20,810,857.25. During this entire period, the landowners never once defaulted in their financial obligation, as is evidenced by the numerous mill rate increases to which they voluntarily agreed. (The above figures do not include the special mill tax rate levied for spruce budworm control.)

The state valuation for assessment of the M.F.D. fire tax was determined every two years by the Board of Equalization. This valuation formed the basis for computation and apportionment of the tax assessed as provided by law. The state tax assessor, by statute, was the agent responsible for determining the amount of annual forest fire tax according to valuations based upon the latest data gained from periodic resurveying of the unorganized territories.*

For those not familiar with the M.F.D., it will be interesting to note that it included fifty-four municipalities, twelve of which are incorporated towns and the remainder plantations. Under the Revised Statute, Title 12, section 1202, any municipality adjacent to an unorganized township and thus to the territory protected by the M.F.D. could at town meeting vote itself into the District with due notice of its action to the state tax assessor, state treasurer, and forest commissioner. However, to withdraw from the District was another matter and required an act of Legislature. This latter requirement has caused consternation in a few municipalities that have made repeated attempts to withdraw during recent sessions of Legislature, but without success.

The main thrust of the municipalities for withdrawing from the M.F.D. was to be included under the "Organized Town" set-up of liability for suppression costs. This meant that the municipality was liable for only one half of one per cent of the state valuation for each town and the state would be responsible for the other one half and also for additional costs above that.

While on the surface this appeared desirable for the municipalities, it was not so for the M.F.D. The District not only suppressed fires, but a lion's share of its budget went into prevention and presuppression. A number of the major corporate owners with holdings in these municipalities, who paid a high percentage of the tax levied by the District, still wanted to remain under the M.F.D. protection program.

The entire M.F.D. program would be endangered by the withdrawal of towns that had no other form of organized fire protection. It was therefore considered unwise to allow a system organized to protect a vast territory to be eroded by carving out small areas on a piecemeal basis. The records show that on the basis of arguments presented, it was not in the best interests of the state to allow withdrawal.

---

* See Appendix II for M.F.D. tax legislative statutes and related tables.

Comparative Financial Statements

1909		1959	
Patrolling Account	$23,096.05	Personal Services	$362,269.85
Lookout Stations and		Professional Fees	995.68
Telephone Lines	7,380.15	Plane Rentals	1,070.03
Watchmen	4,093.54	Plane Operation	7,837.91
Extinguishing Fires	10,944.80	Fire Suppression	34,791.67
Chief Wardens	6,510.99	Travel Expense	8,846.65
Deputy Fire Wardens	1,524.75	Car and Truck Operation	38,331.76
Tools and Supplies	2,163.25	Elec. and Tel. Service	7,259.39
Other Expenses	2,232.32	Rentals	380.20
	$57,945.85	Repairs, Radio	10,738.11
		Repairs, all other	14,116.34
		Insurance	3,411.72
		General Operating Expense	5,776.10
		Food, Tel. and Construction	197.20
		Fuel	1,489.21
		Office Supplies	776.86
		*Uniforms	23.75
		Household supplies and small tools	9,306.85
		Accident Comp. and expenses	781.66
		Retirement Contributions	11,762.50
		Land Purchase	3,230.61
		Buildings and Improvements	15,058.70
		Equipment	69,526.18
		Structures and Improvements	4,294.73
			$612,273.66

*$8,000.00 Budgeted for Uniforms in 1960

Finally a compromise was made whereby under Title 12, section 1601, the dissatisfied municipalities would be reimbursed for fire protection expenditures beyond what the M.F.D. provided up to a maximum of fifty per cent of the District tax paid by the municipality in any one year. Since this enactment, reimbursements have been made of $2,520 for one half of the fiscal year 1971–72 and of $19,862.62 for the fiscal year 1972–73.

An example of the financial impact upon the M.F.D. had these dissatisfied municipalities been allowed to withdraw is clearly shown by the following. In aggregate, the fifty-four municipalities in the M.F.D. paid a fire tax of $198,218.02 in 1971 on a little over a million acres of forest land. This figure represented a substantial revenue for the District's program of forest fire protection. Of course, under the reimbursement plan the District would be liable for a maximum of fifty per cent of $99,000 (one half of the fire tax paid in), but in light of the payments made in reimbursement, it would be some time before any such figure would be reached. In addition, participation in the M.F.D. program acted as an incentive for member municipalities to improve their own fire protection through construction of fire stations and purchase of equipment, and through retainer fees with neighboring towns outside the District by which the use of fire protection equipment could be obtained.

To those funds which the M.F.D. received through the forest fire tax must be added the annual revenues received for fire protection on state-owned lands as follows: Baxter Park, $12,000; Indian Township (Passamaquoddy), $2,040; and State Parks (Allagash and Rangeley), $1,238. All of these areas were within the M.F.D. territory and were assessed at a rate of six cents per acre per year. In 1973, the Legislature changed this procedure to payments for fire protection based upon state-wide annual per acre expenditures for the last fiscal year (Baxter Park, Chapter 87, P.L. 1973; and Indian Township, Chapter 141, P.L. 1973.)

With the passage of the Weeks Law in 1911 (Public Law 435) began a new chapter in the pages of American forest conservation. Now it was possible for the Federal Government to purchase private forest land to incorporate the land into a national forest system. The most important feature of this law was the "granting of federal contributions to the states which organized for forest fire protection."

Maine easily qualified, because of its M.F.D. fire protection program, to receive its proportion of the $200,000 authorization with an annual appropriation of $10,000. She was one of the first three states to receive the initial grant, with substantial increases in succeeding years up until 1924. During this period, these funds were used primarily for salaries of patrolmen and watchmen in the District.

In 1924, the Clarke-McNary Act superseded the Weeks Law and greatly increased fire control activities. This new law was an important factor in helping to reduce the number of fires and the extent of fire damage. Congress authorized $20,000,000 under this law. By 1972, the full authorization was realized, with fifty states participating. Maine received its first federal allotment under this act in 1925.

In 1973, Congress increased the authorization to $40,000,000. It is to be hoped that it will not take another forty-eight years to reach full appropriation of this authorization. States qualifying for funds under the law receive federal allotments based upon a complicated formula of need, expenditures, and acreage to be protected.

While in office as forest commissioner, and as an officer in the National Association of State Foresters, I made several annual trips to Washington, D.C., appearing before the House and Senate subcommittees on Internal Affairs seeking increased funds under the Clarke-McNary Act. It was my pleasant experience to have Senator Margaret Chase Smith of Maine sit in as a guest at these hearings.

In her role as a member of the full Senate Appropriations Committee, Senator Smith was most helpful in getting the recommended fire control increases, as was the late Congressman Clifford T. McIntire, who served on the House Subcommittee on Insular Affairs (agriculture and forestry).

It is important to explain at this juncture that Maine continued to receive federal allotments, and that between the years 1911 to 1920, these amounts were used largely to fund the programs of the M.F.D. Then from 1921 to 1949, as fire protection in unorganized towns improved, a greater proportion of these allotments were diverted from the District. As a result of the 1947 fire disaster and a substantial jump in General Funds, all federal allotments for fire protection under the Clarke-McNary Law were apportioned at the discretion of the forest commissioner on the basis of one third to Organized Towns and two thirds to the M.F.D. This policy continued up to the present.*

In updating the forestry laws for better control and more efficient budgeting, legislation was passed whereby under Title 12, section 513, R.S. 1964, the Forestry Department was designated as the public agency of the state to accept federal, municipal, and private funds in relation to forest fire protection, insect and disease control, management, research and all other matters relating to forests. Although such funds had been received for many years for these cooperative forestry programs, this act provided the proper statutory authority.

All funds received from the District fire tax were placed in a special dedicated revenue account under control of the Department of Finance and Administration. Although not mandatory, this account was administered in a manner to comply as closely as possible with the working procedures of the state in handling its General Fund, but in all cases the fund was administered separately.* Unlike many other state working funds, this account always had the feature of being a "carrying account" so that any unexpended balance did not lapse but continued over to the next fiscal year.**

Preparation of annual financial statements, receipts, and expenditures were on a calendar-year basis until 1958, at which time a change was made to a fiscal year basis in accordance with state reporting procedures. For several years thereafter, the report for the M.F.D. was shown for both calendar and fiscal years. This dual report situation resulted from the District's realization that the landowners wished to see an account of expenditures for a full fire season (April to October).

As we examine and compare the early financial and disbursement

---

* See Appendix II for Maine allocations based on the Weeks and Clarke-McNary laws; also for M.F.D. expenditures 1917, 1927, 1950, 1972.

** See Title 12, Section 1602, R.S. 1964, Maine State Law Library, State House.

statements with more recent ones, we realize the magnitude of change and development that occurred in the fire protection program over the years, leading to a large-scale business operation involving nearly two million dollars by 1972.

One of the early methods of paying forest fire fighters was the use of the so-called "petty cash account." This practice was started during the 1920s and terminated by former Forest Commissioner A. D. Nutting in 1951.

In searching through some old records, it was surprising to find that as early as 1903 Forest Commissioner Edgar E. Ring made a strong recommendation for paying fire fighters from petty cash accounts. In Ring's own words:

> The payment of those fighting fire should be as prompt as possible, in order to encourage persons to respond quickly and cheerfully when summoned to put out a forest fire. In order to obtain this result, the fire warden should be allowed to advance the money whenever possible and to pay the fire fighters immediately after their work is done.

To fully appreciate the importance of prompt payment of wages, one must understand the make-up of the labor force employed in the fighting of forest fires. It was largely composed of French Canadians from Quebec and New Brunswick, who were working in the various pulpwood camps situated in the M.F.D. and operated by the paper companies. These men spoke little if any English. They were recruited from Canada by the various paper companies on a quota basis under a bond of five hundred dollars per person. (This bond has been reduced to seventy-five dollars as of 1972).

When fires started in the cuttings of the employing company, it was difficult at first to get these woods crews to fight the fire. The men argued that their bond covered only the cutting of pulp. I recall how this situation was quickly corrected by a Great Northern Paper Company official, William Hilton, who established a company policy that when pulpwood cutters were recruited, the bonds would stipulate both the cutting of pulp and the fighting of fire. Other paper companies made a similar change in their policy.

Still another source of manpower for fighting fires were men, usually farmers, recruited directly for fire-fighting purposes by jobbers or contractors of the Maine paper companies who had a "following" and knew where to locate such forces. Such recruits went through the U.S. Customs by means of a permit slip or identification card. They were transported directly to and from the fire in cars or trucks by the paper company on whose land the fire occurred. Though immigration officers and the Border Patrol frequently checked Maine camps for violators of legal entry, there was little trouble with these French

Canadians either jumping bond as pulp cutters or as emergency recruits for fire fighting.

When fires burning along the Maine-Quebec line were serious and labor short, the border was sometimes closed. Such a procedure acted to bring men forward to join in the fire fighting, especially when the border closure coincided with weekends. Frankly stated, this was a commandeering of fire fighters, but perfectly legal under existing statutes.

It was not uncommon for the Georgia-Pacific Corporation to bring crews from their woods camps in New Brunswick to fight fires in Maine. Incidentally, there has long been cooperation between the different timberland owners in making manpower and transportation available during major fires. Such cooperation, if for no other reason, was based on the obvious fact that it was better to stop a fire before it spread to one's own holdings.

This type of labor force created the need for a ready and quick method of paying wages aside from the regular payrolls made out at the Augusta office on a monthly basis. Many of the recruited fire fighters bore the same name. Former Chief Warden Kenneth Hinkley of the Rangeley District remembers having nine "Joe Arsenaults" working on the same fire. I have seen a stack of uncanceled pay checks held for lack of proper identification, or equally common, proper address.

The payment by check and sometimes in cash by the supervisor "right on the stump" from petty cash accounts avoided confusion while at the same time promoting good will with the fire fighters, whose services might soon be urgently needed again. A good example of the use and value of such a practice is to be found in the big forest fire in 1934 that started in Quebec and spread into Aroostook County, Maine. The following brief excerpt from a report written by a Canadian International Paper Company forester who was on the scene gives indication of the benefit derived:

> It was learned that the fire fighters employed by the Canadian Government for work on the Canadian side receive less pay and are obliged to wait much longer for their money.... [By paying promptly] Maine Forest Service helped in keeping fire fighters contented and [in obtaining] better results.

Chief Clerk Lillian Coleman of the Augusta office would arrange on short notice with the Depositors Trust to credit five to ten thousand dollars to the checking account of a supervisor to be used in paying off fire fighters. It should be remembered that during these early days company paymasters carried large sums of money into the woods camps so that the men might be paid promptly each month, therefore,

STATE OF MAINE, FORESTRY DIST., H.G.TINGLEY, SUPERVISOR
STATE HOUSE, AUGUSTA, ME.
Examples of bank statements and fire fighters' checks
issued by supervisors from petty cash accounts

Amount Brought Forward			Balance
1.30-		Oct 2 '45	549.29
7.30-	3.00-		
3.00-		Oct 4 '45	535.99
2.00-		Oct 5 '45	533.99
1.60-		Oct 9 '45	532.39
2.80-		Oct 15 '45	529.59
1.30-		Oct 20 '45	528.29
1.50-		Oct 19 '45	526.79

STATE OF MAINE, FORESTRY DIST., G. FAULKNER, SUPERVISOR
STATE HOUSE, AUGUSTA, ME.

Amount Brought Forward			Balance
16.50-		Oct 1 '45	1,313.90
4.00-	2.00-	Oct 2 '45	1,307.90
14.50-		Oct 3 '45	1,293.40
5.00-	5.00-	Oct 4 '45	1,283.40
3.00-	12.00-		
5.00-		Oct 5 '45	1,263.40
6.00-		Oct 6 '45	1,257.40
4.00-	5.96-		
3.00-	5.50-	Oct 15 '45	1,238.94
14.00-		Oct 16 '45	1,224.94
8.00-		Oct 18 '45	1,216.94
42.00-		Oct 24 '45	1,174.94
14.50-		Oct 26 '45	1,160.44
8.75-	5.50-		
12.00-	8.75-		
50.50-		Oct 30 '45	1,074.94

(Unfortunately no cancelled checks are available, for old records were destroyed to provide more storage space.)

The above is an extract from the State Audit Report for fiscal year July 1, 1944 to June 30, 1945.

"Re: Examination of Accounts-Forestry Department and Maine Forestry District Emergency Fire Fighting Fund:

Petty cash funds for the four supervisors on deposit with the Depositors Trust Company, Augusta, Maine, were checked in detail with departmental records, bank statements, etc., and were found to be correct."

in using petty cash funds to pay fire fighters the M.F.D. was following an accepted practice.

As the varied activities of accounting became more complex, not only in the matter of payrolls, contractual services, and inventories, but also in administering the several sources of income in the on-going forestry programs, it was soon apparent that the growing responsibilities of the bookkeeper (later called chief clerk) were too much for one person. This being the case, in 1949–50, Forest Commissioner A. D. Nutting in a cooperative agreement with Finance Commissioner Raymond Mudge, and with the concurrence of Governor Horace Hildreth, included in his biennial budget the new position of a business manager. Mr. Mudge recommended William Whitman, a businessman who had recently retired to Maine from New York, to undertake the position, and he accepted the job. His salary was paid from the state's General Fund.

Mr. Whitman worked out a successful plan for paying fire fighters within twenty-four hours of the receipt of payrolls, excluding weekends. By means of telephone, radio, and plane it was possible to send payrolls into the Augusta office promptly. The supervisors were pleased to be relieved of the responsibility of petty cash accountability, as was the commissioner, who was responsible for checking these accounts returned from the field.

Following the appointment of a business manager, the M.F.D.'s advisory committee appointed working subcommittees on finance and policy. The result was a closer coordination in the preparation of annual District budgets. Previews were frequently made of the forest commissioner's proposed budget by the chairman and the members of his committee on finance. One interesting innovation was the simple narrative statements on certain items which were mailed to all members of the advisory committee prior to its annual meeting. In this way, much time was saved, thus helping to move the budget along toward its final approval.*

The Forestry Department moved into a more complex accounting system in 1949–50. It included tighter budget controls that had been laid down by the Department of Finance and Administration. Such controls involved compliance with federal grants-in-aid, area cost studies, state employment fringe benefits, change-over from unclassified to classified services. At the same time, the financial burdens of the M.F.D. were magnified by the increase in payroll and operation costs.

Principal among the newer accounting procedures was that of "line budgeting." The aim of this concept of line category budgeting (Personnel Services, All Other, and Capital) was to stop the "big boom" spending by department heads during May and June of each year in an attempt to avoid lapsing of funds. In 1953–54, Governor Burton Cross tried unsuccessfully to get the Legislature to adopt this system. At that time department heads were requested to try this method on an honor basis. During Governor Edmund Muskie's term of office, the line budget concept became law under Chapter 130, P.L. 1955 (Maine).

In 1958 field supervisors were trained to prepare their own quarterly allotment work programs and legislative budget requests. It is

---

* See Appendix II for samples of narrative comments in budget reviews for M.F.D. Advisory Committee.

to be remembered that whatever transitions took place, all changes had to be made within the departmental structure and required the approval of the governor and his council.

A previous reference has been made to the cost-conscious attitude in the early days of the M.F.D.'s operation, particularly in the area of laying off personnel during periods of low fire hazard. Under the more recent sophisticated system of accounting, and with the cooperation of the finance subcommittee of the M.F.D.'s advisory committee, this cost-consciousness continued and included many areas other than that of personnel. A few examples follow.

Many towers, living quarters, and storehouses were built with the labor from the fire warden force. Bulk gasoline tanks of one thousand gallon capacity were installed at key warden headquarters. Competitive bids were requested in the purchase of equipment, tools, lumber, and other supplies. Labor and equipment rates were established. Vehicles to be retired from service were sold to realize greater income rather than used for trade-in purposes. A program of phasing out lookout towers in favor of contractual aerial patrols was instigated with considerable saving to the District. Another practice in which large sums were saved was that of the acquisition at little or no cost of Federal Government excess property. Purchase of new light equipment and supplies through Federal Government GSA warehouses at figures one third and often one half the cost on open market greatly expanded the M.F.D. capability at great savings. Finally, the use of company-owned equipment at a rate lower than possible with regular private contractors allowed further economy.

Others areas of cost reduction were explored to determine their feasibility. An illustration would be the study to determine the advantage and saving in contracting for a fleet of vehicles rather than District ownership and operation. This study proved such a plan to be uneconomical due to the large back-country area which had to be covered and the lack of service areas in this territory. However, it can be said without reservation that the effort to save funds was ever present and the results of such economizing became substantial over a period of time.

One item of the budget must be singled out for special mention. This is the matter of the fire reserve account. It was felt that a start should be made to build a cash reserve to meet future emergency contingencies. This idea was not entirely new, for the records of 1911 show that Forest Commissioner Forrest Colby recommended a one-half mill increase in the M.F.D.'s fire tax "to lay up a fund or reserve account against the day which will surely come . . ."

In 1955, the fire reserve account was made a reality and thereafter was kept on a calendar year basis. This was purely an internal,

STATUS OF FIRE RESERVE ACCOUNT (Calendar Year Fire Season)

	Budget	Expenditures	Balance
June 30, 1955			56,752.10
December 31, 1955	$102,000.00	$ 18,249.65	$140,502.45
" 1956	102,000.00	6,556.26	235,946.19
" 1957	102,000.00	27,998.23	309,947.96
" 1958	102,500.00	14,002.14	398,445.82
" 1959	102,500.00	35,861.70	465,084.12
" 1960	102,500.00	140,751.44	426,832.68
" 1961	52,500.00	71,968.31	407,364.37
" 1962	52,500.00	31,884.95	427,979.42
" 1963	52,500.00	59,654.20	420,825.22
" 1964	52,500.00	43,241.66	430,083.56
" 1965	52,500.00	331,334.39	151,249.17
" 1966	52,500.00	65,316.72	
		98,463.23	236,895.68
" 1967	52,500.00	20,956.43	268,439.25
" 1968	52,500.00	117,803.25	203,136.00
" 1969	60,000.00	24,450.85	238,685.15
" 1970	60,000.00	73,434.32	225,250.83
" 1971	60,000.00	30,543.69	254,707.14
" 1972			198,841.96**

*Centerville costs reimbursed from General Fund
**This balance will be kept as a separate account to meet emergencies and amounts will be drawn against it until it is used up. The M.F.D. Advisory Committee has already approved sums to be used for some mechanized fire fighting equipment, capital building construction and budworm control as a fire protection measure. Thus, the balance for 1973 is somewhat reduced. (It eventually was used up.)

departmental bookkeeping procedure. The records in the state comptroller's office showed only a balance of the entire M.F.D. financial program at the end of the fiscal year. Since the District fund was a "carrying account," any balance from the budget item of fire suppression was transferred to the fire reserve, where it was permitted to accumulate from year to year.

At one time the M.F.D.'s advisory committee recommended a ceiling of $500,000 on this reserve fund, but later reduced their recommendation to $300,000. Although this ceiling was never reached, the reserve fund did come close to the latter amount. It should be pointed out that at no time could the forest commissioner tap this reserve without the approval of the M.F.D.'s advisory committee. Funds were used from this account for the replacement of fire losses of equipment, for compensation cases, suppression deficits, purchase of certain pieces of heavy equipment, budworm surveys of potential fire hazard areas, and special capital improvements.

From time to time, several members of the advisory committee expressed concern that the reserve might be suddenly taken away by the state comptroller. These fears were removed by confidence in the forest commissioner's ability to protect this fund and his integrity in regard to its use only for emergency as an *internal* matter of departmental bookkeeping.

Part of the fire reserve account was derived from a 1929 legislative amendment to the Maine Forestry District Law (Chapter 152, section 66), which read as follows:

> ... upon receipt of information from the forest commissioner that there is in said fund a certain sum in excess of the amount necessary for the protection of the forests in said district from fire, the governor and council may issue a warrant to the treasurer of the state to refund proportionally to the landowners paying the tax assessed as aforesaid, such sum or sums as shall be recommended by the forest commissioner.

This legislation was no doubt prompted by the fact that despite the disastrous fire season of 1921 all M.F.D. indebtedness to the state and the landowners had been paid, and by January 1, 1930, there was a net surplus of $172,945.67.

Acting under the 1929 legislation, Forest Commissioner Neil L. Violette wrote a letter advising all landowners within the territory protected by the M.F.D. that he was recommending a refund of $50,695.28 through a thirty per cent rebate on the next forest fire tax. Both the governor and the council passed the following order: "That the State Treasurer be authorized to rebate the 1930 Forest District Tax by thirty (30) per cent." Two more rebates were made in the years 1932 and 1934, resulting in refunds of $54,401 (thirty per cent) and $31,775.64 (twenty per cent).

The practice of rebating was an unprecedented move and was discontinued after 1934. From then on any surplus was permitted to accumulate, finally resulting in 1955 in the formal creation of the fire reserve account.

The entire policy of the Forestry Department, and of the M.F.D. in particular, was based upon the concept of "adequate protection." The question has often arisen as to exactly what was meant by this phrase. Simply stated, adequate forest fire protection means the utilization of all available funds, manpower, and equipment to handle the average, normal fire situation. Costs of a program geared to meet the occasional "blow-up" fire situation would be prohibitive.

The adage that the most expensive fire can be the cheapest if no effort is used to control the conflagration still holds true today. For example, a ten-acre fire that costs $20,000 can be considered a cheap fire based on what might have happened if the $20,000 had not been spent in an all-out suppression effort to prevent from further loss the forest resource values of timber, water, recreation, scenic beauty, and wild life.

Arriving at a figure of cost per acre per year for forest fire protection in Maine involves several factors. Some indication can be made from the area and cost studies conducted every five years by the U.S. Forest Service for purposes of allocating federal funds. These results are based upon the factors of need, acres of forest to be protected against fire, and expenditures.

Area and cost studies were begun through a U.S. Senate Resolution in 1920, as part of an extension of the 1911 Weeks Act. This study led to the passage of the Clarke-McNary Act of 1924. The latter act did not limit fire control expenditures to forested watersheds of navigable streams, as had the Weeks Act. From that time on, surveys of fire protection costs and needs have been made at intervals of five to seven years. Consistently obeyed since the passage of the Clarke-McNary Act is the requirement that the federal allotment to any state will not exceed one-half of the current estimated cost of the total protection job.

The most recent study under way as of this writing in 1974 will be the tenth. Instead of being labeled an "Area and Cost Study," it will be known as a "Forest Protection Analysis." Its specific objectives "are to enable the states and the U.S. Forest Service to evaluate the character and size of the fire protection job in these days of rapid change and rising costs, to appraise relative progress, and to serve as a basis for regular allotments to the states on an equitable basis."

These studies have served as a goal for reducing forest fire destruction to a figure called an "allowable burn." The figure for Maine at one time was one tenth of one per cent of the forest area or about 18,000 acres. Acreage burned under this figure was considered a good year. With improvements in fire-fighting techniques, intensified protection, and the good record in Maine, the "allowable burn" is now five hundreths of one per cent, or about 9,000 acres. Maine has been well within or under this figure for many years.

The cost of forest fire protection per acre per year has increased progressively. In the broadest sense, it revolves around the multiple use of forest resources and the risk factors tied in with expenditures. Records show early estimates of three and six cents per acre, but as cost studies became more accurate, protection costs increased from eleven cents in 1960 to eighteen cents per acre per year in 1972 for the state's total of 18 million acres. No separate statistics have been worked up just for the M.F.D., but it is believed that they are comparable to the state-wide cost.

VI

A TESTIMONY TO COOPERATION

> *Owning over one million acres of timberland is a responsibility. Our goal is to accomplish the best utilization of our renewable fiber resources while we act as good stewards of the land.**

The history of the M.F.D. is an account of a program dedicated to the protection of the living treasure of the Maine forest. It is also a testimony to the cooperation between public and private interests that gave birth to the program and nurtured its development.

While much of this chapter will deal with the cooperation of the landowners with the forest fire protection programs of the M.F.D., it is only right to pay tribute to the dedication and cooperation that has marked the office and field forces of the organization itself. The following quote from a letter written by Forest Commissioner Neil Violette to his chief wardens well expresses this *esprit de corps*.

> Results will show this year that the Maine Forestry District has made a remarkable record in forest fire protection, which we believe is due to a great extent to your splendid service and hearty cooperation. Being financially able and in view of the length and condition of the season, we wish to show you our appreciation and that of the landowners by enclosing herewith a bonus, which is small but [we] sincerely hope that it will be received in the same spirit in which it is given.

The very extent of the territory within the protection of the M.F.D. posed a difficulty in the establishment of a closely knit organi-

---

* Morris Wing, Regional Manager of Woodlands in Maine, International Paper Company, Jay.

zation, particularly in earlier years before the advent of better systems of communication, as an anecdote from the recollections of former Forest Commissioner Samuel T. Dana illustrates:

> Since I enjoyed field work, I probably saw more of the field force than most of my predecessors. On one trip I talked for some time with a lookout watchman before disclosing my identity. When I finally did so, he exclaimed, "You, the Forest Commissioner? My God, I thought you were a gray-haired old son-of-a-bitch." Several years were still to elapse before that became an accurate description.

Despite the vast reaches of the M.F.D.'s wilderness territory and the increasing bureaucratic responsibilities, there was a constant effort on the part of the top echelon in the Augusta office to keep closely in touch with the field force, not only through the supervisors and chief wardens but in person.* There was a common bond between all personnel, and that was the forest itself and its protection, which demanded united action.

The cooperation of the landowners has been well illustrated in relationship to the formation of the M.F.D. and the subsequent support of this organization through the self-imposed District fire tax. However, there were numerous other ways in which landowners assisted in M.F.D. programs.

A little known contribution on the part of private companies was the funding of one-fourth and half-page advertisements in the daily newspapers, weeklies, and magazines stressing the importance of fire prevention during periods of high fire danger. Prominent coverage has been given to Smokey Bear, class fire danger boards and the "Keep Maine Green" program accompanied with appropriately worded warning messages. Ads and feature stories also have appeared in special newspaper supplements on industrial and recreation issues. At no time has the M.F.D. budgeted or paid for this advertising.

The industry and landowners made a substantial cash contribution when the Maine Forestry Department hosted the National Association of State Foresters' ninety-seventh annual convention in 1969. Also there were generous donations of wood and paper products for each visiting state and federal forester, along with special door prizes. The comment commonly heard was "How fortunate for Maine to have such a good working relationship between the forest commissioner and the wood-using industry and landowners."

In the plane owned by the International Paper Company, the

---

* As forest commissioner, I visited over fifty M.F.D. lookout towers, walking the long telephone lines and trails to camps and towers as part of our policy of bringing the Augusta office to the men in the field.

pilot and the Woodlands Regional Manager, Morris Wing, have on numerous occasions assisted in a real reconnaissance of forest areas in northern Maine during critically dry times by giving "on the spot" information during actual fire situations. Private planes, chartered by other companies, have also been very helpful in detection work.

During the period when chief wardens were only seasonally employed by the District, the landowners often provided winter work. Such jobs varied but included marking timber, cruising, checking camps, and scaling. When spring came, these wardens went back into the District payrolls. This kind of cooperation served the dual purpose of providing year-round employment and building morale within the M.F.D. warden force until more permanent positions were created.

Most helpful have been the dollar a year leases made by the landowners to the M.F.D. for the sites of lookout towers, warden living quarters, storehouses, campsites, and right-of-way for telephone lines.

Training schools have been one of the most important activities in the M.F.D. fire prevention program. These classes were not only for the fire wardens but also included the personnel of the private companies and landowners engaged in various woods operations. Often officials from the Maine offices of the companies would either participate in or attend banquet get-togethers.

The recently organized Maine Logging Road Committee of the Seven Islands Land Company, the International Paper Company, Oxford Paper Company, Prentiss and Carlisle, and the Great Northern Paper Company, is another example of cooperation with the M.F.D. In this case fire wardens were issued special passes or were permitted to go through check points on personal recognizance. The distribution of these passes to outsiders is strictly limited for obvious control purposes.

Another important item in the working relationship between the forest commissioner and the landowners was their support of his biennial report and of special forestry oriented bills brought before the Legislature. Such assistance was often in the form of appearances at public hearings or through letters to the chairmen of key committees. Similar support came on requests for increases in federal grants-in-aid in fire control and other related forestry legislation.

Cooperation was a most important factor in the actual prevention and fighting of forest fires.

The Great Northern Paper Company made it a policy to provide portable fire tool houses and trucks loaded with fire-fighting equipment on all their woods operations. Other companies have a similar arrangement with their wood contractors.

A final example in this brief list of illustrations of mutual assistance expands the scope of cooperation to include not only Maine landowners but also the Province of Quebec's Forest Protection Service.

In 1934, a series of brush fires started by Quebec homesteaders were burning under the bad conditions of a high wind and dry soil cover. These fires originated on the Canadian side and spread into Maine with one fire covering approximately 60,000 acres. Supervisor George Gruhn of the M.F.D. flew over the fire area in a plane based at Quebec City, but was prevented any good aerial view of the fire's perimeter due to the dense smoke.

The Augusta office advised the Canadian International Paper Company to take any steps necessary for recruiting fire fighters, with payment to be made from the M.F.D. account. Help was also promised to move two entire CCC camps from another part of the state to the scene of the fire.

The following is a quote from the report of a Canadian International Paper Company forester:

> Mr. Gruhn insisted that transportation on above equipment both to and from Daaquam be billed to the Maine Forest Service. It is however here recommended that in consideration of the many courtesies in behalf of the Company, including aerial observations, etc., any charges other than their shortages in equipment be held to a minimum of bare cost, if not entirely overlooked.

The annual rash of spring fires started from the clearing of land by settlers along the border were a deep concern both to Maine landowners and to the M.F.D. Finally, Mr. Keiffer of the Quebec Forest Protection Service and the Maine Forestry District officials worked out a plan of mutual cooperation. This was a prelude to international mutual assistance which, years later, resulted in the Northeastern Forest Fire Protection Compact for Canadian participation.

The Northeastern Forest Fire Protection Commission deserves special consideration as an example of wide-range cooperation. Maine, as one of the nine member states and provinces (the six New England States, New York, Quebec and New Brunswick), has been a very active participant in the affairs of this commission.

The need for mutual aid and cooperative structure of such magnitude became apparent during the disastrous forest fires of the autumn of 1947 that struck all of New England, and Maine in particular. The State of Maine alone experienced the greatest devastation, with 220,000 acres of forest land burned, 2,500 people made homeless, nine communities leveled or practically wiped out, $7,000,000 sustained in timber damage, and $300,000 in fire suppression costs. Throughout New England the experience pointed out the critical importance of interstate and federal cooperation to deal effectively and promptly with such situations.

As a consequence of these events the Eighty-first Congress on June 25, 1949 enacted Public Law 129 authorizing formation of the Northeastern Forest Fire Protection Commission. Three years later, on May 13, 1952, the Eighty-second Congress enacted Public Law 340 authorizing Canadian participation in the compact.*

With this congressional authorization, the six New England States and New York ratified their joinder action through their respective legislatures during the years 1949–1950. Joinder action followed nine years later on September 23, 1969, by the Province of Quebec, and on June 9, 1970, by the Province of New Brunswick.*

At the time of these actions by the Canadian provinces, I was privileged to be chairman of the Commission and with Governor Curtis of Maine represented the New England Governor's Conference, and participated in the very colorful signing ceremonies at Quebec City, Quebec, and at Fredericton, New Brunswick.

It is of interest to point out that Maine, as a member state, has common international boundaries amounting to two hundred and ninety-three miles on the Quebec side and three hundred and nineteen miles on the New Brunswick side. As much of this boundary distance runs through forest, this fact was of particular importance to the M.F.D.'s program of fire protection and prevention.

Maine pays the highest annual assessment of $3,572 based upon a formula having a base of $400 with the remainder prorated upon acreage and average expenditures taken over a five-year period. The total annual assessment budget and acreage under the compact protection are $14,000 and 54,452,000 acres respectively. The compact provides for mutual aid assistance, trained overhead crews, and equipment in the event of a major forest fire situation. There is a constant updating of fire plans. In addition, there are the continuing committees on policy, equipment, legislation, and technical and executive training. Maine has invoked mutual assistance on four occasions since joining in the compact.

Maine personnel from the Forestry Department have been active in the organizational structure of the compact. I served as its chairman for two years (1968–1969); and the present Director of the Bureau of Forestry, Fred E. Holt, served most effectively for many years as chairman of the training committee.

In 1972 the Northeastern Forest Fire Protection Commission commemorated its twenty-fifth anniversary. As a fitting tribute to the service of this compact, Maine's Senator Edmund S. Muskie read into

---

* Governor Horace Hildreth of Maine, through the New England Governors Conference, was largely responsible for getting the Council of State governments to draft suggested congressional legislation for a forest fire compact.

COMPACT SIGNING CEREMONY AT QUEBEC CITY, QUEBEC, SEPTEMBER 23, 1969
Above, from left to right: Forest Commissioner Austin H. Wilkins, Maine; p
Minister of Natural Resources Claude G. Gosselin; page
Below, foreground, from left to right: Minister of Natural Resources Claude G. Goss
Quebec; Forest Commissioner Austin H. Wilkins, Maine; Governor of Maine Ken
M. Curtis; Prime Minister of Quebec Jean-Jacques Bertrand

COMPACT SIGNING CEREMONY, FREDERICTON, NEW BRUNSWICK, JUNE 9, 1970
Above: New England and New York Compact Commissioners attending the ceremony. *Below, left to right:* Natural Resources Minister William Duffey, New Brunswick; Governor Kenneth M. Curtis, Maine; Forest Commissioner Austin H. Wilkins, Maine; Premier Louis Robichaud, New Brunswick

INTERSTATE FOREST FIRE PROTECTION COMPACT

In Witness Whereof

I, FREDERICK G. PAYNE,

Governor of the State of MAINE,

have set my hand for and on behalf of the State of

MAINE and affixed the Seal of said

State this 23rd day of August, A. D. 19 49

"Pursuant to Chapter 75, P & S Laws of 1949"

Governor

ATTEST:

Deputy Secretary of State

the *Congressional Record* a full chronology of events that had taken place during the organization's history. This documentation was prepared jointly by the U.S. Forest Service and members of the Commission.*

The general interest in forest fire protection has involved many agencies and groups of people over the years.

The State Police on numerous occasions have added another communication channel through use of their radios during fire-fighting operations. Such assistance has been of particular value in those fires involving Organized Towns. Their assistance has also been valuable in handling traffic and setting up road blocks to keep sightseers out of fire areas. In cases where hose lines have been laid across highways, traffic control by the State Police has been essential. On occasion police escort has been of great aid in moving heavy mechanized equipment over the highways.

Civil Defense has assisted in making available their mobile field kitchens for the serving of hot lunches to fire fighters. Through their network of communications, they have helped to mobilize men and equipment when requested.

The State Highway Department has also assisted in granting permits for movement of bulldozers and trucks over highways. From their regional field headquarters certain types of mechanized equipment have been made available upon request and in several instances used. During severe periods of ongoing fires, former Commissioner David Stevens of the Department of Transportation would contact the Forestry office, either by person or by telephone, to offer whatever assistance possible. On one fire he wrote off the cost as a contribution toward reducing excessive suppression expenditures.

In the earlier days of passenger railroad service, railroad companies have transported trainloads of men to fire locations. In other cases the employees of various industries have been made available for fire fighting. In 1911, during a fire in Township A, Range 11, the American Thread Mill at Milo shut down its operation, sending its entire crew by Bangor & Aroostook train to a point of disembarkment close to the fire location.

Years ago, during one particularly dry period when numerous fires were burning, commercial airlines deviated fifty miles from their usual flight lines to patrol. They reported fires into the nearest airport, and the location was then relayed to the Forestry Department's radio dispatcher in Augusta.

The U.S. Forest Service has been most cooperative. Requests for assistance have been promptly met, both in the form of data supplied

---

* See Appendix III for Foreword to the Commission's Reference Manual.

INTERSTATE FOREST FIRE PROTECTION COMPACT

IN WITNESS WHEREOF

I, _____Claude-G. Gosselin_____ Minister of Lands and Forests of the Province of __Quebec__, accept membership of the Province of __Quebec__ in the Northeastern Forest Fire Protection Commission and set my hand for and on behalf of said Province and affix its Seal this __23rd__ day of __September__, A. D. 19__69__, pursuant to Order in Council No. 2497 of the Quebec Government, dated August 27th, 1969, a copy of said Order in Council being attached hereto.

Claude G. Gosselin
Minister

Accepted on behalf of the Northeastern
Forest Fire Protection Commission

Austin H. Wilkins
Chairman

CONVENTION ENTRE DIVERS ETATS RELATIVEMENT À
LA PROTECTION DES FORÊTS CONTRE LE FEU

EN FOI DE QUOI

Je, _____Claude-G. Gosselin_____, Ministre des Terres et Forêts de la Province de _____Québec_____, accepte la participation de la Province de _____Québec_____ comme membre de la "Northeastern Forest Fire Protection Commission" et appose, au nom de la Province de _____Québec_____, ma signature ainsi que le sceau de ladite Province ce _____23ième_____ jour de _____septembre_____ en l'an de grâce 19_69_, conformément à l'Arrêté en Conseil No. 2497 du Gouvernement du Québec, en date du 27 août 1969, copie dudit Arrêté en Conseil étant annexée aux présentes.

Claude G. Gosselin
Ministre

Accepté au nom de la "Northeastern Forest Fire Protection Commission"

Austin H. Wilkins
Président

Northeastern Forest Fire Protection Compact

IN WITNESS WHEREOF

I, William R. Duffie, Minister of Natural Resources of the Province of New Brunswick, accept membership of the Province of New Brunswick in the Northeastern Forest Fire Protection Commission and set my hand for and on behalf of said Province and affix its Seal this 9th day of June, A.D. 1970, pursuant to Order in Council No. 70-231 of the New Brunswick Government, dated April 8th, 1970, a copy of said Order in Council being attached hereto.

Minister

Premier
Province of
New Brunswick

Accepted on behalf
of the Northeastern
Forest Fire Protection
Commission

Chairman
N.E.F.F.P.C.

Governor – Maine,
Chairman
N.E. Gov. Conference

NEW BRUNSWICK

and in the form of materials. Frequent visits have been made by personnel of this federal service to the Augusta office as well as in the field. The latter visits have been more a matter of service than of perfunctory inspections.

A special salute goes to the press and to radio and television stations that have cooperated to the fullest in their coverage of forest fire situations, class danger days, woods ban proclamations, and other related aspects of forest fire protection.

From before the formation of the M.F.D., when landowners paid out of their pockets for fire protection, sometimes in full or through cost sharing with county commissioners and later with the state, there has been a growing awareness that the forest is both a public and a private resource and that it is the business of all to preserve this treasure.

In the legislative act of 1909 that created the M.F.D., there was no provision for an advisory committee. The basic intent was to give full power to the forest commissioner. It was his duty alone to establish and maintain for the first time an organized forest control program in the unorganized territory.

However, as added responsibility grew in administering the District, and especially as matters pertaining to financial affairs grew more complex, the need for an informal advisory committee to assist and council the commissioner became more and more evident.

This committee started as a so-called "finance committee," chosen by the landowners during the latter part of the tenure of Forest Commissioner Raymond E. Rendall (1942–1947), for the purpose of closely examining his request for an increased District tax. Another factor leading to the creation of an advisory committee was Governor Horace A. Hildreth's insistence on letters from several of the large landowners and their representatives promising that they would ask the next Legislature for a tax increase to cover a loan that had been made from the state's surplus. In 1948, the District had borrowed funds on the personal guarantee of the governor to the state comptroller. (It should be noted that this obligation was met.)

As a result of all this, Rendall's successor, A.D. Nutting, appointed the first M.F.D. advisory committee to the commissioner in 1948. It represented a cross-section of landowners, both in terms of extent of land owned and as to type of ownership.

Its membership changed from time to time over the years; at last count (1972) there were fourteen members, representing corporate ownerships of industry and large private land managers. Acting under the committee were two helpful subcommittees dealing with finance and policy.

Given below are the members of the advisory committee for the years 1948, 1959, and 1972.

FIRST M.F.D. ADVISORY COMMITTEE: 1948
A.D. Nutting, Forest Commissioner: Chairman-Secretary

Bradford, Grover C.	Pingree Timberlands, Bangor
Burns, Kenneth	S. D. Warren Paper Co., Westbrook
Freedman, Louis J.	Penobscot Development Co., Old Town
Herr, Clarence S.	Brown Company, Berlin (N.H.)
Hilton, William	Great Northern Paper Co., Bangor
Pearson, Frank	Eastern Pulpwood Co., Calais
Philbrick, William	Coburn Heirs, Skowhegan
Pierce, James M.	Madigan and Pierce, Houlton
Sawyer, George C.	Dunn Timberlands, Ashland

M.F.D. ADVISORY COMMITTEE: 1959
(50th Anniversary of the District)

Blaisdell, George M.	International Paper Company, Chisholm
Bradford, Grover C.	Pingree Timberlands, Bangor
Crocker, Floyd M.	St. Regis Paper Company, Bucksport
Demeritt, Dwight B.	Standard Packing Corp., Brewer
Ellis, G. Donald	Scott Paper Company, Winslow
Herr, Clarence C.	Brown Company, Berlin (N.H.)
Hilton, William	Great Northern Paper Company, Bangor
Merrill, Robert W.	Penobscot Development Co., Old Town
Philbrick, William	Coburn Heirs, Skowhegan
Sawyer, George C.	Dunn Timberlands, Ashland

LAST M.F.D. ADVISORY COMMITTEE: 1972
Chairman Morris Wing

Bork, John H.	Brown Company, Berlin (N.H.)
Carlisle, George D.	Prentiss & Carlisle Co., Inc., Bangor
Currier, Ralph	Great Northern Paper Company, Millinocket
Hartranft, John L.	Oxford Paper Company, Rumford
Mitchell, Roger J.	Georgia-Pacific Corporation, Woodland
Philbrick, William	Coburn Heirs, Skowhegan
Sawyer, George C.	Dunn Timberlands, Ashland
Semonite, David	J. M. Huber Corporation, Portland
Sinclair, John G.	Seven Islands Land Company, Bangor
Stedman, Arthur F.	Scott Paper Company, Winslow
Weller, Herbert, J.	St. Regis Paper Company, Bucksport

Williams, Niles C.	Dead River Company, Bangor
Wing, Morris R.	International Paper Company, Chisholm
Wood, Raymond J.	Diamond International Corp., Old Town

Subcommittee—Finances	*Subcommittee—Future Trends*
Chairman Arthur F. Stedman	Chairman John G. Sinclair
Carlisle, George D.	Sawyer, George C.
Hartranft, John L.	Williams, Niles C.
Wing, Morris R.	Wood, Raymond S.

Austin H. Wilkins, Forest Commissioner—Secretary

The advisory committee had no legal status. Meetings were called at the discretion of the forest commissioner. The agenda was prepared by him and the meetings presided over by the chairman of the committee. All meetings were informal, with notes kept by the commissioner; no official minutes were recorded.

In addition to the annual meetings to review the budget and other pertinent business matters, occasional special meetings were called to discuss raising of the District tax, emergency taxation to fund the effort against budworm outbreaks, radio conversion from low to high frequencies, and for the ten-year periodic review of the Maine timber resources. At such meetings annual labor and equipment rates were also established.

It should be made clear that at no time did this committee attempt to function as a policy making body or to usurp the powers of the commissioner. At one time, some consideration was given to formalizing the committee through legislative action. However, since the working relationship was good and the District fire tax income continued to be a dedicated revenue and in no way involved with the state's General Fund appropriations, it was thought best to keep the function of the committee and its informal nature as originally designed.

Under the new Tree Growth Tax Law (Chapter 616 P.L. 1972) and the Conservation Department Act of 1973, it might well be worth considering the establishment again of an advisory committee, but on a larger scale, to include the total forest resources of the state. Incidentally, it was John Sinclair of the Seven Islands Land Company who, while serving on the subcommittee on future trends, initiated legislation that resulted in the Tree Growth Tax Law.

Whatever the future brings, it is to be hoped that the same spirit of cooperation will persist in a united effort to protect the resources of the forests of Maine.

Old wooden tower, Depot Mountain, 1909

VII

WATCHMEN AND TELEPHONE LINES

In our present American way of life, does forestry look attractive? Men engaged in it say it is a good thing because they value the experience that touches at so many points.*

Much has already been said about the vastness of the territory under the protectorate of the M.F.D. Faced with a duty of surveillance on such a scale and with the obvious necessity of being able to spread alarms rapidly, and equally important, to dispatch fire-fighting forces in time, it is not surprising to find that the history of the M.F.D. reflects at every stage a search for better means of fire detection and alarm.

Even as the first steps were being taken to organize the M.F.D., private landowners were setting up the first lookout towers from which one man might scan a multitude of acres. They were, incidentally, establishing a "first-time" record in the country.

At the turn of the century, America had no forest fire detection and suppression system as we know it today. Forest fires burned across the country, destroying huge acreages of timberland annually. Few and untrained hands were lifted against this insidious enemy, and little, if any, thought was given to detecting fires while they were still small.

Stimulated into action by the severe fires of 1903, several Maine timberland owners, operators, and other interested people began to give their attention to the recurring problem. Names such as Elmer

---
* Austin Cary, nationally known pioneer forester, from a paper presented in 1916.

Crowley, William Shaw, W. J. Lannigan, and Payson Viles made history at this point.*

Elmer Crowley's own story of the first lookout tower, built on Squaw Mountain, makes interesting reading:

> I was employed as forest engineer by the M. G. Shaw Lumber Company, whose woods operations were handled by William Shaw. I was employed for one year beginning in June 1904.
>
> I arrived in Greenville soon after graduating from the University of Maine. The next day I went to a lumber camp on the south side of Big Squaw Mountain Township. A summer logging operation was then being carried on by means of an overhead cable way, which I understood was the first to be used in this vicinity. After looking the machine over, we traveled northward along the tableland of Big Squaw Mountain lying easterly of the summit. On our return trip, Mr. Shaw suggested that we go to the top of the mountain, which we did. I will never forget the impression that this view made on me, it being my first trip to the top of a mountain of any considerable size. . . . Mr. Shaw pointed out the various objects which we could see . . . Milliken Farm, the Corner Farm, Indian Pond, Moxie Mountain and Pond . . . Shirley mill with smoke visibly issuing from the sawmill stack. Twenty-six miles to the south we could see smoke coming from the sawmill stack at Guilford. At Greenville we could see another sawmill, and while sitting there we noticed a train coming through Misery Gore. . . .
>
> It was then that the thought occurred to me and on the instant I asked Mr. Shaw if this would not be a good place for a forest fire watchman. I expressed my opinion that one man on the mountain could do more and better work from this point of vantage than 100 men traveling through the woods . . . in ½ hour a man could get down to the nearest telephone and get word down to Greenville. The only remark that Mr. Shaw made . . . was that he thought a man hurrying down the mountain to to report a fire would probably break his neck before he reached the phone.
>
> The subject was not further brought up until late in the summer. I was then working on a plan for improvement to the skidding machine. Mr. Shaw came into the drafting room and instead of looking at the plan as he often did . . . he stepped a few feet away and said nothing. I looked at him and noticed a

---

* See Appendix IV for a definitive fire occurrence table for Maine in 1903–1972.

quizzical expression on his face. He said, "Crowley, I don't know but that was a pretty good idea." Naturally I inquired what he was talking about and he replied, "Putting a man on Squaw Mountain to watch for forest fires. . . ." *

That was the beginning.

The lookout was placed in operation on June 10, 1905 with William Hilton of Greenville, now Vice President of the Great Northern Paper Company, Bangor, as the first observer, or "watchman" as he was called at that time.

The first entry in the log kept by Mr. Hilton reads: "Commenced work Saturday, June 10, 1905; clear, South wind." Hilton served as observer from 1905 through 1908. During the first year he lived at the M. G. Shaw logging camp, making the trip up and down the mountain each day.

The value of the Squaw Mountain lookout was demonstrated many times in 1905. Two other towers were built by private funds that year, on the bald tops of Attean and Bigelow mountains. Between 1905 and 1908 six more were built by landowners, and in 1909 all lookout stations in the unorganized towns and plantations were absorbed into the Maine Forestry District upon its creation by the Legislature.

Former Forest Commissioner A.D. Nutting in 1958 inaugurated forestry field days in recognition of various aspects of forestry. It is significant that the first two field days commemorated important dates in the history of the M.F.D.

On July 26, 1958, a field day was held in Greenville, Maine, to commemorate the erection of the tower on Squaw and to pay tribute to its first watchman. Mr. Dwight B. Demeritt, Standard Packaging Corp., of Brewer, Maine, was master of ceremonies. The principal speaker was Richard E. McArdle, Chief of the U.S. Forest Service, Washington, D.C. A paper prepared by James L. Madden, Vice President of the Scott Paper Company, Chester, Pennsylvania, entitled, "How Our Forests Are Used," was read by Mr. Donald Ellis, Scott Paper Company, Waterville, Maine. Forest Commissioner A. D. Nutting spoke on the Maine Forestry District. Mr. Louis J. Freedman, retired Vice President of Penobscot Development Company, presented a certificate of recognition to Mr. Hilton for his "pioneering in forest fire control."

An interesting part of the afternoon program was this radio message from the tower:

---

* Quoted from Report of Forest Commissioner 1939–1940, pp. 85–86. See also Appendix IV for notes from Elmer Crowley's Diary.

Greetings Bill Hilton and Louis Oakes. This is station KCB 428 atop Squaw Mountain where in 1905 you, Bill, began your career as Maine's first forest fire lookout reporting down the mountain to Louis Oakes at Greenville, first state forest fire warden in this area.

Fifty-three years have passed into history since you climbed this famous mountain to open the first of more than 6,000 forest fire lookouts now operating in the nation.

The trail to the summit is much easier now than when you made the daily trip up and down from M. G. Shaw's logging camp. The view is just as beautiful and the expanse of wilderness still reaches as far as the eye can see. And lightning still has this peak as its frequent target—a habit that sent you scurrying down the mountain to safety several times each summer!

Some things have changed, however. The crude alidade you used has been replaced by instruments of greater precision, and the patchwork map of the forest land under your care has been replaced by one of more accuracy.

Perhaps the most remarkable change has been the perfection of radio as a means of prompt communication with headquarters. During your employment as observer your reports of forest fires were telephoned to Louis Oakes at Greenville. Mr. Oakes, who is with you today, certainly recalls some of the exasperating delays and other difficulties of those early days. Today with the miracle of radio, vital information is transmitted with ease and speed. What a help that would have been to you both in 1905!

Many other changes have taken place, of course. Many amusing and exciting incidents will be recalled to you both today. Perhaps some of them occurred on this very peak. If so, I am sure the group would enjoy hearing of them.

And now, from atop of Squaw Mountain, the site of the nation's first forest fire lookout, we return you to Squaw Mountain Inn and the exercises commemorating the event.

For posterity a bronze plaque embedded in a granite boulder is located on a turn-out just off the highway, in full view of Squaw Mountain and the lookout tower.

The evolution of lookout tower construction, maintenance, and communication in the back country during the early days is sometimes hard to believe, as is the fact that the system functioned as well as it did. This is especially true when one compares the conditions then with those of today with our easy access into wilderness country, our many conveniences that make both living and working easier, our advances in mapping techniques, and improvement in protective de-

Above: William Hilton, left, first lookout observer on Squaw Mountain, talking with first chief forest fire warden (1909) Louis Oakes.
At right: present chief fire warden John Smith
Below: Bronze tablet commemorating Squaw Mountain lookout tower. First continuously operated tower in the country, erected in 1905, at Greenville, Maine

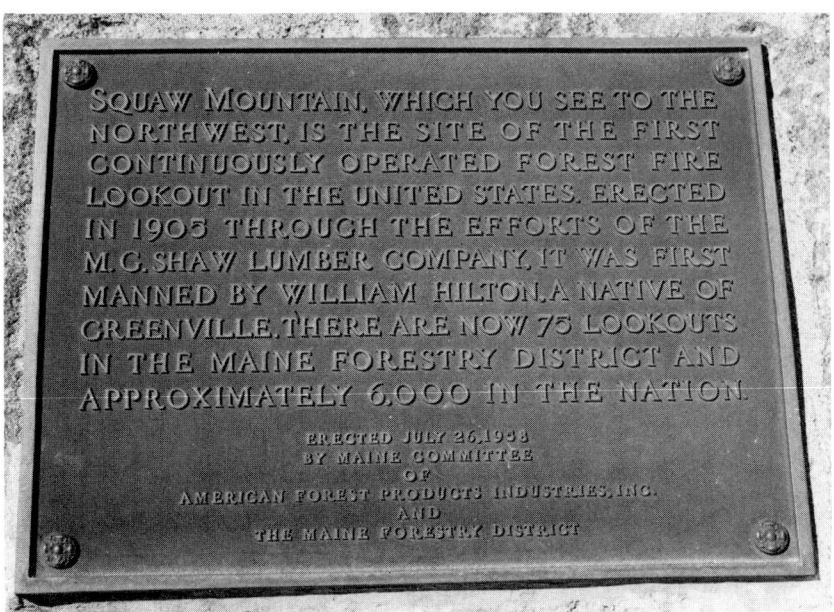

Closeup of tablet inscription

vices against lightning—not a small consideration on top of Maine's higher mountains.

It was a special breed of men who manned the remote lookout towers. In the early days they moved into their camp and tower by walking or by tote team and did not come out until fall. They were usually good woodsmen, hardy and dependable, with a knowledge of the surrounding countryside. They were neat housekeepers, whose wood-burning stoves would shine to the envy of any housewife. They did their own cooking, sewing, camp repairs; cut their own fuel wood, and caught rain in barrels from the gutters at each corner of the camp for their washing.

Some kept small gardens and maintained vegetable cellars along with preserves of fiddleheads, wild raspberries, strawberries, blueberries, and cranberries. Those fortunate to have wives did considerable canning. It was necessary to maintain high fences to keep out the deer. Marauding bears were another matter. Broken cross-cut saw blades were used to guard the windows against the raiding of these animals in their search for easy food.

Supplies were toted in and left in specially made boxes at the foot of the mountain to be back-packed up the steep trails by the watchman. Attempts were made in later years to supply the watchmen via "free fall" and parachute drops, but this method did not prove very successful.

Left: Unused wooden tower on Old Washington Bald (1918) replaced by new structure (1934) T42 M.D. Washington County; *right:* an early lookout

Lightning was a constant hazard, often driving the watchman pell-mell from his high tower and putting his lines of communication out of commission. Lightning has been known to come in on the wires, shriveling telephone mouthpieces and receivers to the size of a pencil. In spite of grounding, these incidents continued to occur. There have been many direct hits on towers and camps.

Many of the early watchman camps were made of peeled logs. With mice and snakes as indoor companions, bottom logs often had to be replaced. Nearly always some repair job was waiting for a rainy day. Later, when logging operations opened up the country, giving better access by team and truck, the old peeled log camps gave way to those built from dimension lumber.

The department started to publish a series of handbooks in 1962, the first of which was the *Forest Watchman's Handbook*. The following quote is reproduced to illustrate how vitally important the role of the watchman was in the first line of surveillance for the M.F.D.

You, as a forest watchman, are often referred to as the eyes of the Maine Forest Service. Upon your performance hinges the speed with which ground crews can be alerted and dispatched to a going fire. You often give first and only warning of fires in remote areas. With the increased role of radio communication in the fire detection and control organization, you may be called upon to perform a dual role in detection and communications.

The lookout tower system is the oldest means of smoke detection and is still the most commonly used. As a forest watchman, you are a vital part of this system.

Above: Early log watchmen's camps
Below: Modern framed watchman's camp

Always remember that to be a good watchman, you must know the country you see over, and you must make your fire reports quickly and accurately. Everything you can learn about your area will be to your advantage and will in turn increase your value to the Service. Take every opportunity to get around in your territory. Make every effort to thoroughly study your seen area. Correlate this first-hand knowledge with your maps. Your efficiency in locating fires will be increased proportionately.

The Maine Forest Service, and especially your co-workers and associates are depending on you. You are a part of the team effort in fire control. Although you may sometimes feel like a forgotten man, alone on your mountain top, remember that as a forest watchman, you perform a key job in the protection of the forests of Maine.

Throughout the era of lookout towers, the evolution of their construction displayed many forms. Early towers were invariably made of logs placed to form a quadruped or a tripod topped with an open platform. Then came log cribbed-cabins and finally the more sophisticated towers, standing on poles but completely enclosed with boards and shingles. The wooden structures gave way to steel, still topped with wooden cabs—the first being constructed by the M.F.D. in 1913. Later the cabs were also fabricated with steel, as were the stairways and landings.

The following quote gives an interesting glimpse of early towers:

In 1915 we had for use in our work a lookout station on Mt. Chase, one on Horse Mountain and one at Beetle Mountain. The lookout on Chase was made of boards and had board windows. This contraption was wired down on the bald top of the mountain. At Horse Mountain we had a scaffold eight or ten feet high made of four poles. At Beetle we had a log cabin with board door and windows. We had four telephones but only two of them were working. We had no maps in the lookout stations and so had to guess at the location of a fire.

The usual cost-consciousness characteristic of the M.F.D. shows itself in the construction of towers. Steel towers on Washington, Bald, Cooper, and Wesley mountains were erected from salvaged one-fourth-inch angle iron from the U.S. Naval radio tower that had been located at Bar Harbor, Maine.

Old-time wardens can recall the hardship of early days—swamping out a trail or road for a packhorse or team to tote the materials for erecting those lookout towers. Then there were endless hours of hand drilling holes into the ledge for insertion of the eye bolts that were to

EARLY CRUDE WOODEN LOOKOUT TOWERS
Below, left: a tree serves for support of Round Mountain lookout

Upper left: early log cribbed wood structure towers; *at right:* all frame wood towers; *lower left:* wood cab on steel frame

Some of the old solidly enclosed towers

Steel towers with wooden cabins, erected in 1917

secure the steel legs and guy cables. Once in, the steel and eye bolts were made solid by pouring melted lead or brimstone (sulphur) into the holes. Proper anchorage was essential if the towers were to withstand the strong winds of summer and the severe ice and snowstorms of winter. Ice has been known to "rime-on,"* creating tons of extra weight and crushing the wooden cabs or even bending the entire tower structure over.

During the latter years of the lookout tower era, came the jeep and truck and finally the most modern method of airlifting by helicopter, all of which made erection and construction easier. As commissioner I was privileged to observe at close hand two projects in which a department helicopter airlifted pre-cut sections for replacing old wood cabs with steel frames. Ironically, the coming of such modern

---

* The term used for the formation of layers of hoar-frost.

Results of severe weather

Steel tower with wooden cab   Latest all-steel tower

techniques was the harbinger of the end of the primary importance of the watchman and the network of mountaintop towers.

The growth and decline of the M.F.D.'s lookout tower system presents an interesting story. Researching through the files, one would estimate that over one hundred such lookouts were erected between 1905 and 1973. During this period there was much relocating; many towers were abandoned due to the movement of lumbering operations to new areas; many new ones built to fill in blind spots as an overall effort was made to establish some uniformity and adequate coverage of all the vast area within the District.

After the first spurt of tower building in 1905–1908, another came in 1913–1914 when seventeen more towers were built, prompted by additional District money and the ability to pay wages of the watchmen from federal money provided by the Weeks Law of 1911. (It is a

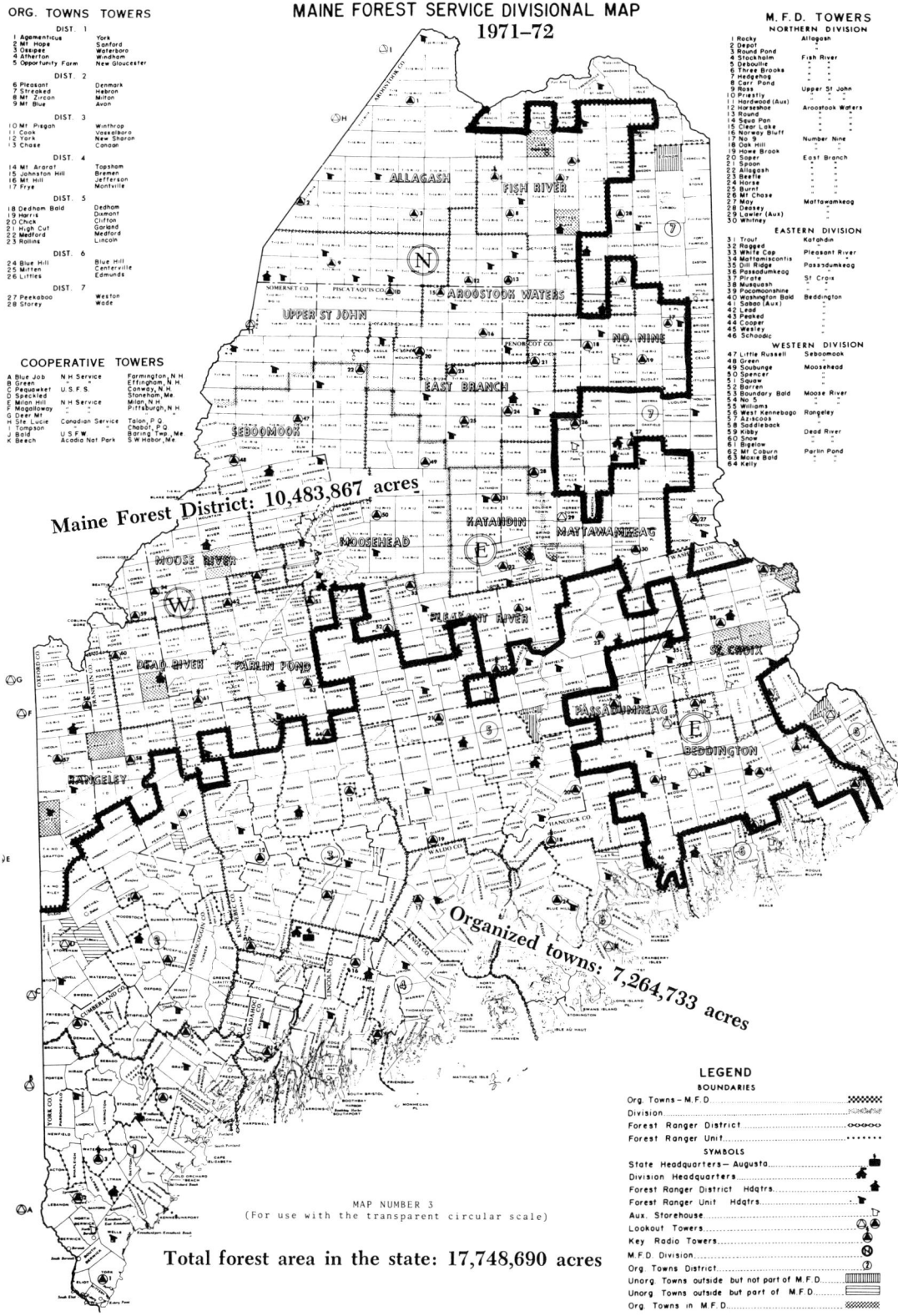

Pilot George Johnson and I landing on top of Mt. Bigelow. Flight time seven minutes base to summit; return trip four minutes

little known fact that at one time some of the towers were manned by watchmen under federal pay.) Between 1916 and 1925, twenty-two more towers went up on the mountaintops. Later, an average of a new tower per year was added to the total, so that by 1935 there were seventy-seven in the M.F.D., the peak number of towers in use. From a standpoint of construction cost, it is interesting to note that while the earliest towers cost some $750, the later steel towers cost from $10,000 to $15,000, depending upon the size and location.

The year 1950 saw the beginning of the decline of operational towers, the number decreasing to a total of fifty-nine by 1962. It should be mentioned that some towers were retained on an auxiliary basis and could be reactivated during extremely dry forest conditions, but the number of active towers diminished rapidly with the advent of aircraft surveillance, until by 1973 the number stood at thirteen.*

---

* See Appendix IV for lists of lookout stations operated in 1917, 1932, 1943, 1953, 1973, and older, abandoned stations.

MT. BIGELOW LOOKOUT TOWER, ERECTED IN 1917
Height: 16½ foot steel frame and 7 foot high wood cab. In 1961–62 the tower was replaced by a fieldstone base and sturdy wood cabin to withstand severe winter weather

In nearly all cases, the landowners were most cooperative in the erection of the towers. One exception was the proposed tower on Borestone Mountain where conditions laid down by the landowner were too severe. The tower was erected on top of Barren Mountain instead.

In contrast, an excellent example of cooperation and assistance from landowners is the case of the tower on top of Kineo Mountain. Originally a peeled pole camp covered with tar paper, the lookout was replaced in 1917–18 with a steel tower and wooden cab. Funds came from the Kennebec Protective Association, the Maine Central Railroad, and the Ricker Hotel Company.

A number of lookout towers served outside the perimeters of the M.F.D. Those on Rocky Mountain (1907), Depot Mountain (1909), and Hardwood Mountain (1916) looked directly into Canada. An earlier reference has been made to the brush fires of Canadian homesteaders that on occasion spread into Maine. The Maine lookout watchmen found a special language problem on such occasions when it became necessary to communicate across the border in the process of locating a fire. It was not uncommon to have three hundred acres of brush fires going at the same time on one hundred parcels of land.

There were also cooperative lookout towers involving the Maine-New Hampshire border, the Maine-Quebec border, as well as those in the organized towns and those involved with Acadia National Park at Bar Harbor and the U.S. Wildlife Moosehorn area in Baring and Edmunds townships.

The forest commissioner, as the chief agent in the cooperative venture to build and improve the forest protection system, had the authority to select those towers which would be operated as federal lookout stations. As payment for his services in this role, he received one dollar per month. Watchmen received two dollars and fifty cents per day and were paid at the end of each month by federal check upon vouchers submitted by the forest commissioner. These watchmen were in essence federal employees and could return to their towers each season.

A letter dated September 16, 1915, from Forest Commissioner Frank E. Mace to chief wardens, served retroactive notice that the Federal Government would take over certain payrolls in that year:

> The pay roll of *watchmen* and *patrolmen* will be taken over by the Federal Department beginning September 1st 1915.
>
> It is very essential that the bills of the watchmen and patrolmen for the present month should be in this office, properly approved by the chief wardens, by September 25th.
>
> If it is likely that the men are to be kept on to the end of the month—make out their bills to and including *September 30th.* If there is any change in any of the time after bills have been sent, wire this office not later than September 30th.
>
> Careful attention to these orders will result in prompt payment from the United States Treasury.

Note: Funds available from Weeks Law 1911.

There is no known record of the number of federally paid watchmen for the year 1911, but the expenditures under the federal allotment of $10,000 for that year are given as $9,986. From available figures, there were twenty-two federal tower watchmen each year for the period 1917 to 1920. In 1920, twenty-eight patrolmen were designated as federally paid from funds generated by the Weeks Law.

As the fire protection system grew, both in the M.F.D. and in the Organized Towns, these allotments under the Weeks Law increased, but when the Clarke-McNary Act was passed in 1924, Maine's allot-

A remote tower, guardian of the forest

Inside a typical watchman's tower cabin, closeup of topographical map table

ments became a budgetary matter under state procedures, and the watchmen and patrolmen were no longer paid directly with federal checks.

The equipment furnished to watchmen steadily improved. Early crude alidades, patchwork maps, old-fashioned battery-operated telephones with their hand cranks, and low-power binoculars gave way to modern range finders, precise maps, latest radio sets, and high-power binoculars.

Improvements made in tower maps are of special interest. Originally a watchman had little to go on in locating and reporting fires. Much depended upon his knowledge of the area, combined with guesswork and the help of hand-drawn maps. Then, in 1917, an ingenious mapping instrument called a "relief" or "panorama alidade" was first used in Washington County to prepare tower maps. It was the design of Frank H. Coburn, of New Hampshire, who held the copyright. The project in Maine was a joint cooperative effort between E. S. Atkinson of the U.S. Forest Service and Archie Norcross of the Maine Forestry Department.

The process involved in plotting a profile of the surrounding country as seen from a lookout tower is as follows:

A heavy piece of drawing paper is placed upon a map stand. From its center a 15-inch radius is drawn, leaving an outside margin of 3 inches. The interior of the circle is reserved for filling in a plane topographic map of the surrounding country for horizontal control. The pivot point of the alidade is then placed in a hole in the center of the mapping board. Its forward arm extends out to the 3-inch margin. After adjusting the forward sight so that the front sight covers the highest mountain peak upon the horizon and at the same time the needle point falls within the outer edge of the paper, the rear sight is adjusted to include all the foreground possible and still cover the highest peak. The front sight is then adjusted to coincide with the change of contour along the mountain peaks, ridges and water lines. By turning a small crank it is then possible to pin-plot the entire surrounding country in a 360° profile.

It usually takes three to four good clear days to complete a panorama from a single tower. There are only certain times of the day when the profile mapper can get all the details of a given area without interference of shadows, clouds, and other factors. Thus, there is considerable moving around from section to section with the alidade before the entire profile contour is completed.

When the work is done on the heavy piece of drafting paper, it is then taken to the Augusta office for winter work and final mapping. Supervisor Robert Stubbs of the Western Division did a considerable number of tower profiles and I also assisted in some of the tower field and office work. The relief or panorama alidade we used is now a museum piece.

The following steps were taken in the office for final completion of a tower map. All the field data from the panorama 3-inch wide circle on the drawing paper was transferred to a tracing cloth. This information was checked and double checked from all available references, then the interior of the 15-inch radius circle was carefully filled in, forming a plane topographic map. It is important that the location of a mountaintop, especially one which has a neighboring tower, lines up accurately with the mountain peak showing on the panorama profile. Only in this way can there be a successful triangulation.

All M.F.D. towers in time were provided with such maps. These were ideal for "crossing off" or triangulating on forest fires. Within the 30-inch diameter map were locations of other neighboring towers and by checking with each watchman's azimuth it was possible to pinpoint the fire location. All maps were on magnetic north.

Later the panorama profile map became obsolete and a start was made toward "grid map systems." Not too many M.F.D. maps were converted, since the lookout tower soon gave way to aerial detection. Thus passed into history another interesting "first" in Maine for the M.F.D.

As important as finding the location of fires was the means by which such vital information could be quickly communicated and coordinated. The story of the establishment of a telephone network by the M.F.D. is as interesting as it was vital to the forest protection program.

The M.F.D. telephone system was both elaborate and intricate. Nowhere else has there been a similar system of woods telephones tying together towers spread across 10,000,000 acres of unorganized territory with all their challenges and problems. Nor was the use of this system of communication, or the sophisticated radio network which was to follow, limited to forest-fire protection. Services were also provided for special military exercises, emergency calls, search and rescue operations, as well as other errands of mercy.

In 1905 the woods telephone system began within the unorganized territory of Maine, when the three lookout towers on Squaw, Attean, and Bigelow mountains were erected with ground lines leading into lumber camps or offices. These first lines of communication, like the towers themselves, were paid for by the landowners.

Then in 1909, with the creation of the M.F.D., came the rapid expansion of a fire protection system, with the resultant growth of the telephone system into a giant pattern of spiderlike webs for each of the four geographical divisions of the District.

The peak development of the M.F.D.'s telephone network was reached in the early 1950s with a total of approximately 3,500 miles of ground and metallic circuit lines. After this period, many of the lines were gradually abandoned in proportion to the increase of implementation of radios. Today, there are about one hundred miles of woods telephone lines left in operation.

Division supervisors Robert Hutton, Robert Stubbs, Harry Tingley, Rex Gilpatrick, and George Faulkner must be credited with perfecting the excellent networks within their respective areas. With the help of patrolmen, watchmen, and chief wardens, all lines were kept in remarkable condition, and they handled a large volume of traffic.

Construction and maintenance of such a vast communication network in a wilderness territory was a challenge. Construction called for careful planning, and thousands of man hours were spent in keeping the undergrowth clear of the lines. In the reports written by forest commissioners, one frequently finds references to hundreds of miles of lines cleaned or bushed-out, new lines built and lines replaced—a never-ending fight to maintain efficient communication.

Problems involving steep slopes, open ledges, and rocky mountaintops, bogs, streams, and river crossings, as well as the crossing of cutover areas had to be overcome. Severe snow, ice and windstorms played havoc with the lines. Each spring all lines were checked by patrolmen walking on snowshoes over the crust of deep snows. Broken branches and fallen trees had to be removed, and in many cases whole sections of lines had to be cut out and replaced. High winds caused considerable damage, especially during that period when there was a heavy tree mortality resulting from the bronze birch borer (die-back). It was not uncommon, during the regular fire season, for watchmen and patrolmen to find it necessary to leave their stations in order to clear lines from fallen branches after high winds.

Moose also caused a problem. These animals were known to get entangled with low-hanging wires and to walk away, tearing off a quarter to a half mile of wire, which was never found. In other instances, and especially during the spring patrol, moose were found strangled to death or dead from exhaustion in their effort to free themselves from the wire. There are records of over a fourth mile of wire found wrapped around the antlers of one of these unfortunate animals.

Still another problem was the proper grounding of telephone lines against lightning. There were many instances of lines and telephone sets being completely knocked out of service by severe electrical storms. Long arcs of fire and sparks would come in on the wires. Many watchmen could relate some harrowing experiences with such storms.

There was also the matter of logistics in obtaining equipment and

TRAGEDY IN THE WOODS
Above: One-eleventh of a mile of tangled woods telephone wire was wrapped around this rack of moosehorns during the animal's struggle to get away
Below: Moose strangled to death by M.F.D. woods telephone line

materials—hundreds of barrels of split porcelain insulators packed in sawdust, hundreds of cases of glass insulators, strings of wooded brackets, miles of number ten and twelve galvanized iron wire in half-mile rolls, many coils of double-twisted, covered lead-in wire, hundreds of wall telephone sets, cases of dry cell batteries, and hundreds of pounds of staples. There were also lineman's tools, belts, climbing irons with straps and pads, various types of pliers for cutting and splicing wire, field-test boxes, canvas bags for carrying insulators, etc. In addition parts, such as switches, sleeves, coils, and ringers, had to be stockpiled.

All that remains today of the accumulation are a few scattered wall telephone sets, which have become collector items valued as high as two hundred dollars each, and the abandoned lines left in the woods.

Training played a big part in the maintenance program. The warden force had to be knowledgeable in the repairing of telephone sets, splicing lines, proper methods of grounding, switchboard hook-ups, field-testing procedures, and many other phases of work. To the credit of the warden force are many innovations that made for easing the work load and improving the network.

At the annual warden training sessions, held in the spring, the use and operation of the telephone system was often made a major part of the program. Many wardens will remember "how-so-ever" Johnson of New Brunswick, the telephone expert who visited Maine several times and participated at these sessions. Bus Tingley, son of Harry Tingley, assistant supervisor in the M.F.D.'s northern region, was among the experts in telephone work and most helpful. It should be noted that representatives of the various landowners also attended the sessions since many of the lines had hook-ups with lumber camps.

At one time, the District carried on its payrolls the title of "lineman." His duty was to serve as a troubleshooter, visiting areas in the various divisions with the object of improving the woods line telephone or of solving particular problems.

In the course of time, there were occasions for change. In some areas lines went into commercial telephone central offices, in which case it became necessary to convert from a ground circuit (single wire) to metallic circuits (two-wire system). This was especially so when the dial system came into being. In other instances where the District's lines paralleled electric power lines, conversion to metallic circuits was necessary because of the noise factor. Adjustments also had to be made for connecting the M.F.D.'s system to the many miles of private lines belonging to lumbering operations and camps. Such connections were essential, for such operators were sources of manpower and equipment. As the public utility systems expanded, there were miles of District ground lines placed on poles leased from these companies.

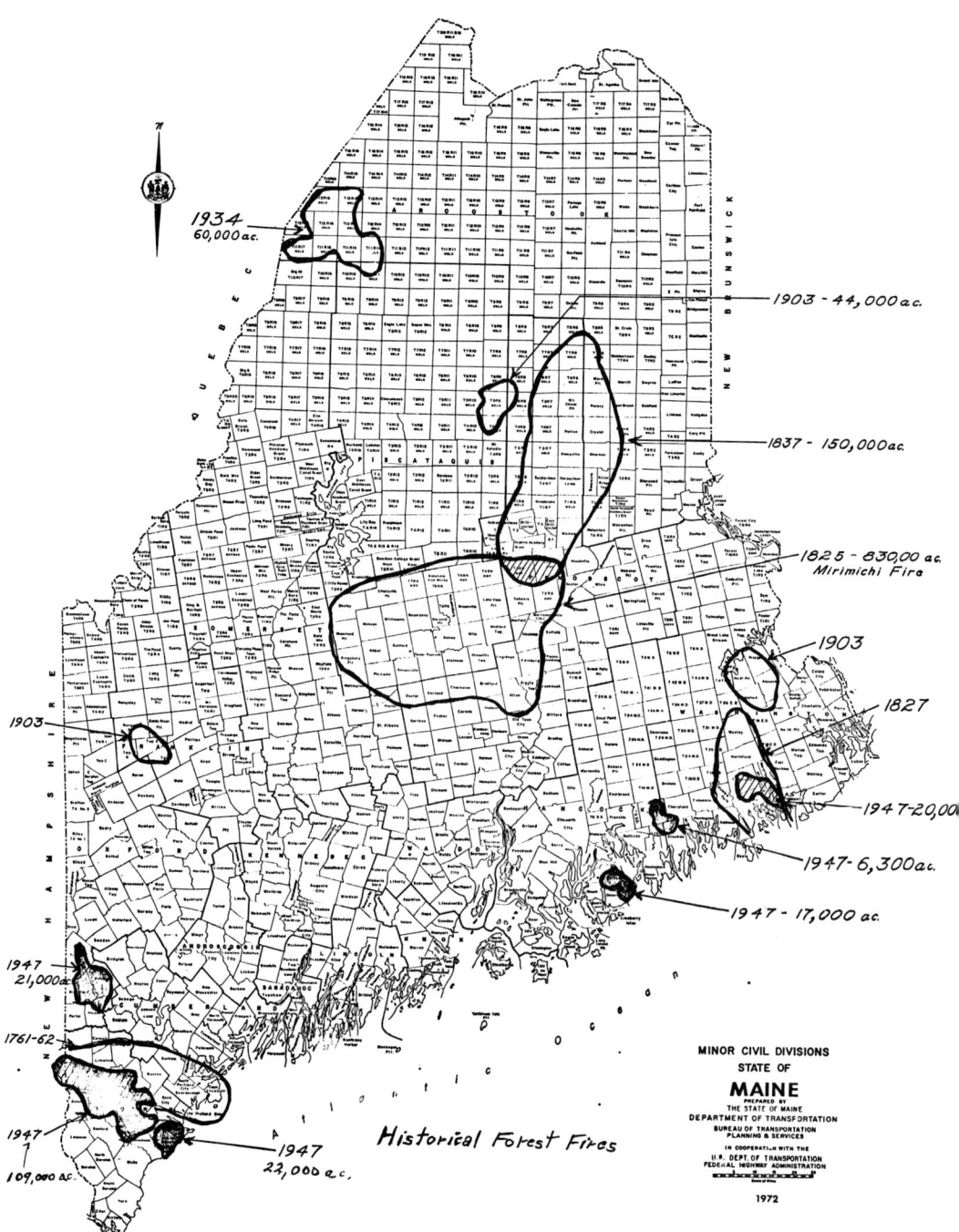

VIII

FIRE SUPPRESSION AND REPORTING

*To understand the forest and to realize that a resource can support a variety of uses is an important fundamental in these days when we place so much emphasis on the environment.**

Having given proper attention to the watchman atop his mountain tower and to the miles of woods telephone lines that carried the reports of his eyes to the proper dispatching centers, we now turn to the reason for all this wide-flung surveillance and means of communication—the awesome sight of fire and the suppression of this great destroyer of the green forest.

Unfortunately for the purposes of forest fire prevention, the public too quickly forgets those major conflagrations that have turned huge tracts into blackened wastelands. One need not search his memory far back, however, to recall the fall of 1947 when all in Maine were made conscious of the threat of that appalling scourge of fire running wild and unchecked. The "yellow days" of June 1941 may also come to mind—those days when a huge cloud of smoke drifted over the state carried at a high altitude from big fires in the Province of Quebec. The odor of burning wood was noticeable for several days before the pall actually settled over Maine. Visibility was so low during the days that followed that airplane travel was stopped and lookout towers had but limited range of view over the forest they guarded.

What the public may too soon forget has always been first in the minds of the personnel of the M.F.D. In the matters of fire prevention

---

* John Sinclair, President Seven Islands Company, Bangor. (Quoted from *Maine High Adventure Area and Boy Scouts of America Log Book.*)

Horse teams hauling fire equipment on a drag into a fire before the days of roads and bulldozers

and fire suppression, the history of the M.F.D. again records a story of constant improvement.

The following letter of June 11, 1915, from Forest Commissioner Mace to John Mitchell, chief warden at Patten, illustrates the concern and attention to detail that he practiced, which was typical of M.F.D. commissioners:

> With regard to the telephone conversation that you had with the Deputy, I am glad to learn that you have a crew of men on this fire and trust that the same is under control by this time. In the future, do not for a moment think of letting any fire burn simply because it is on waste land. Get a crew on it at once as this is the only way to successfully control a forest fire without any great amount of damage.
>
> With regard to the Trout Brook Station, will you kindly inform me the circumstances, how this burned. Was there a camp there also and did that burn up. I have commissioned no one for this place and do not know whether you had a man there or not. You will take steps to re-build a log tower and also please let me know whom you are going to place there. As to the patrolman on the stage road, if you think that it is absolutely necessary, kindly recommend a man or I might have someone to send you.

An amazing change has occurred in the last forty-five years in the design and development of both hand tools and power equipment used in forest fire fighting. In the early days, standard equipment consisted of the common pail, a gunnysack, a bough or branch, a single or double bit axe, and the shovel.

A most interesting suggestion for an addition to this equipment is found in a letter written by Chief Warden John Mitchell in 1918, in which he proposed that the District have at least one hundred sap-carrier yokes made and distributed. His letter explains his request:

> I was notified of a fire on Wissattaquoik Lake on T4, R10. It burned over 8–10 acres. It started where the big Pogey fire started on June the 2nd, 1915. This fire burned in the turf which was composed of rotten wood and punk, burning from one inch to two feet deep amongst the rocks . . . the only way to successfully fight this fire was by carrying water. This gave me the idea of the sap yokes.

Records do not show whether Mitchell got his yokes, but Louis Oakes, of Greenville, another chief warden in the earliest days of the M.F.D., in giving his reminiscences recalled that sap-carrying yokes were tried, but proved to be too cumbersome for woods travel.

In a letter written in 1918 by Forest Commissioner Forrest Colby to a chief warden, the commissioner states, "I have also ordered sent to you a dozen canvas buckets which I think will be enough to start with."

I once interviewed former Chief Warden Blin Page, who recalled using boughs and gunnysacks and carrying water pails in fighting a fire in Holeb Township. What a contrast to modern methods and present equipment!

Irving G. Stetson, writing to me in 1964, made the following comments on early fire fighting and the costs involved:

> By the way, I recall the big Lobster Mountain fire which occurred in July, 1911, during the four year period when I was cutting logs on the East Middlesex Canal Township. I hasten to say that the fire was not caused by my crew, as I had no men there after the drive left Lobster Lake early in June, but was caused by lightning. At any rate, I got 50 men and fought the fire for ten days. At that time the state—presumably the Forestry District—maintained a stock of rather primitive tools, judged by present standards, at Northeast Carry, which we used to fight the fire. Our books here show that the Agents' account paid me $768.29 on November 14, 1911, to cover my expense, and that on November 10, 1913, we eventually wangled $359.85 out of the State Treasurer in partial reimbursement of the $768.29. If the

Fire on Lobster Mountain during the summer of 1911. A rare photo of an early ongoing forest fire in Maine

fire had occurred in 1908, I presume that our owners would have had to pay the whole bill. It would be rather interesting to compare what fighting that fire cost in 1911 with what it would probably cost these days, probably around $6,000 to $7,000; so perhaps the M.F.D. tax rate of $.007 is not so exorbitant after all!

The following bills for supplies sold to the Forestry Department and used in fighting a fire in 1915 illustrate the fact that fires were fought by hand and that the manpower wielding the simple tools had to be fed.

The letter from Mr. G. E. Hyde, of the Eastern Manufacturing Company of Bangor, that accompanied the bill, dated 11/4/1915 and addressed to John Mitchell, Fire Warden of Matagamon, follows:

Enclosed please find a bill which we had rendered to the State of Maine Forestry Department, to which we have received a reply from Frank E. Mace, Forest Commissioner, stating that all other items aside from the $1.67 charged to you are to be paid by the men themselves to whom the supplies were furnished. Will you kindly take up the matter with the several individuals and collect same, and forward the proceeds to us, and for your trouble in doing so, of course we could cancel the charge to yourself.

SUPPLIES DELIVERED —

To Andrew Finnegan, fire warden at Webster Lake Dam:

1	Gal. K. Oil,	.20	
3	lbs. Soda,	.21	
3	" Cream Tartar,	1.50	
1	Bar Soap.	.05	
5	lbs. Table Salt,	.25	
¼	" Ginger,	.10	
¼	" Pepper,	.12	
5	" Sugar,	.40	
3	" Pork,	.51	
1	" Tea,	.30	
1	Pk. Potatoes,	.25	
3	Qts. Beans,	.33	
25	Lbs. Flour,	1.25	
1	Pr. Socks,	.75	6.22

To James Cody, fire warden — Hauling Canoe
        Second Lake to Pine Knoll,   4.00

To John Mitchell at Swing Camp,

May 10	6 Meals,	1.20	
	½ Bu. Oats	.47	1.67

To Frank Brown, fire warden at Swing Camp:

June 18 – 1 Qt. K. Oil,		.05	
3 Bu. Oats,		2.82	2.87

To Boynton & Brown, fire wardens at Swing Camp:*

2	lbs.	Cream Tartar,	1.00	
2	"	Soda,	.14	
25	"	Flour,	1.25	
10	"	Sugar,	.80	
10	"	Tea,	3.00	
2	"	Pork,	.34	
2	"	Lard,	.36	
			6.89	3.45
				$18.21

---

* One-half of this bill was paid by Boynton—the other half is due from Frank Brown.

About the time of the First World War, a whole series of specially designed types of hand fire tools were manufactured and became available in large quantities. These included light shovels with round points, fire rakes, grub hoes, cutter mattocks, Pulaski fire axes, brush hooks, and the "Indian" back-pack pump.

In a recent report the M.F.D. still carried on inventory a total of over 12,000 hand fire tools for use in building hand-dug fire lines. In the days when large one-to-two-hundred-men lumber-pulp camps and CCC camps were a source of manpower for the fighting of fires, the number of hand tools was much greater. But the hand tool is still important. There is an old saying that it is the foot soldier who holds the conquered land. Likewise, it is the fire fighter who patrols on foot with shovel, axe, or back-pack pump who holds the line after the fire has been brought under control.

Following the era of almost exclusive use of hand fire tools came the portable power pumps, relay tanks, one-and-a-half-inch hose and, finally, power saws. Emphasis was on portability and lightweight equipment, including pumpers, canvas relay and hose packs, and linen hose.

Some of the older wardens and landowners will recall the old model portable pumps of Northern, Evinrude, and Fairbanks-Morse. Later came the Gorman-Rupp, Briggs and Stratton and, more recently, the high speed, gear pumps (3,500 rpm) of the Homelite & Pacific models. In contrast to the older models, which required two-man carrying racks, these later models were of such weight that they could be attached to pack boards for back-packing to the fire site. Records show that by 1972 the M.F.D. had approximately 800,000 feet of rubber and linen hose, three hundred and sixty-four portable pumps with accessories, and two hundred and two relay and portable tanks.

Still another development in forest fire fighting equipment was the introduction of heavy bulldozers, tank trucks, trailers, fire line plows, and trucks of various types. The Forestry Department at one time carried an inventory of three hundred and sixty-seven cars and trucks, most of which were in the M.F.D. That number placed the Department as second largest among all the state agencies in terms of vehicles operated.

The emphasis upon mechanization in the suppression of fires within the State of Maine undoubtedly can be attributed to the 1947 fire disaster and the fires of 1952. In that year, there were three hundred and one fires; 18,615 acres were burned (mostly within unorganized territory) with losses or damages of $535,899 and with suppression costs of $439,532. Following such experiences came the introduction of new tactics and techniques in the establishment of fire lines in which the mechanized equipment replaced the use of hand tools

Fire equipment flown in by plane and operating from shoreline

whenever possible. Seldom does a fire occur today without one or several heavy mechanized units appearing on the scene.

Only fire wardens can fully appreciate the struggle, often against the toughest of odds, in getting water onto a forest fire. Monumental changes have occurred since the early days of carting water in pails and canvas buckets. In those first days of the M.F.D.'s fight against fire there were cases where bucket brigades were the only expedient. Later came the period when the effort, often frustrated, was centered on relaying water through long lines of hose through the use of portable pumps and relay barrels or canvas and "Harodike" bags. It was not uncommon to relay water through long lines of hose using four to six power pumps. Lifting water onto fires on mountain slopes required a special technique to assure the proper synchronization of these pumps, for if one pump failed, the whole effort was lost or temporarily delayed. Sometimes these pumps were hooked up in tandem and water pushed through lines over one thousand feet in length. A novel idea was used in the 1952 Misery Gore fire. Two-and-a-half-inch sections of threaded iron pipe were used along with a trailer pumper to transport water over one half to three quarters of a mile. Later, aluminum irrigation pipe was used effectively for the same purpose on some fires. More recent has been the use of 500 to 1,000-gallon capacity tank trucks to supply the water. The increased use of air attack by water-dropping

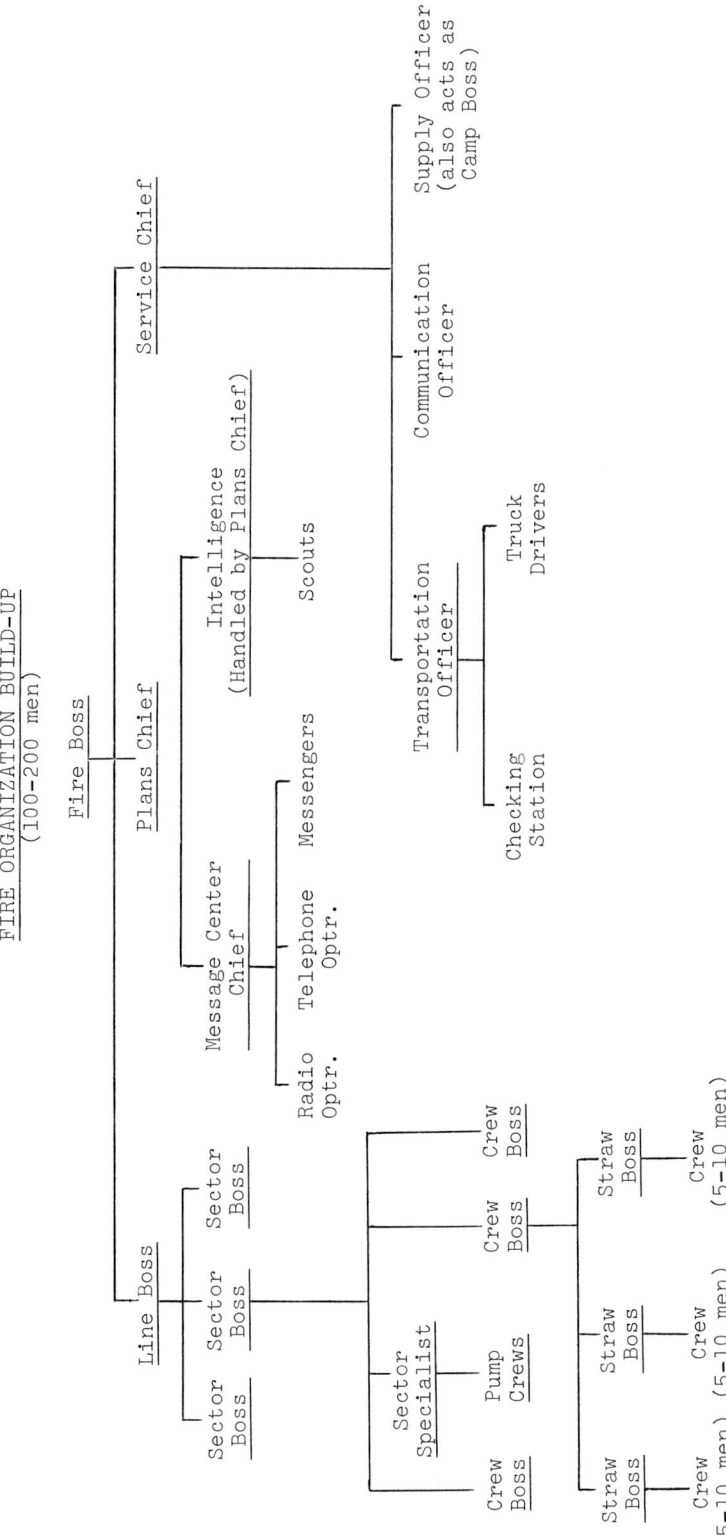

128

and spreading of chemical retardants should be mentioned here as the latest advance in the fighting of forest fires.

The increase in mechanization was accompanied by an increase in organizational structure in the field personnel involved in fire fighting.

In the building of the M.F.D.'s inventory of equipment, special credit should be given to the Federal Excess Property Program. This program started in 1957. The Maine Forestry Department, as a state conservation agency, was eligible to receive and acquire excess property for forest fire protection purposes. The Department was very active in screening catalogs of such excess property and in the inspection of various armed service bases or depots in New England and New York for items that would be useful in the M.F.D. program.

At one time nearly fifty-five per cent of the Department's fleet of vehicles originated from federal excess property. Such acquired equipment included four-wheel drive vehicles, fire trucks, bulldozers, house and tractor trailers, helicopters, Beaver planes, etc. These were acquired gratis or at costs rarely over twenty per cent of the original price.

Between 1960 and 1964, the Department acquired $1,432,631 worth of property under the excess program. Items acquired in one year valued at $360,000. Costs would have been prohibitive budget-wise had this equipment been purchased on the open market.

This method of acquiring equipment, however, had its drawbacks. The time came when the cost of upkeep on such equipment proved a liability. After that the vehicles were kept licensed, but were placed upon a stand-by basis only.

To the equipment owned and operated by the M.F.D. must be added that made available by landowners during the time of fire.

During the rapid evolution of forest fire-fighting equipment, the M.F.D. has attempted to keep informed of new developments. Through contacts with the U.S. Forest Service and its testing laboratories, as well as through the ingenuity of the M.F.D.'s wardens and mechanics, there has been a continuing program of research and field testing in the search for more efficient equipment.

The total aggregate of present mechanized power for fighting fire makes a striking contrast to the effort and means available during the first decades of the twentieth century.*

A significant economic factor is involved between mechanization and the cost of labor in fire control. The once great source of manpower to be found in Maine lumber and pulp camps no longer exists.

---

* See Appendix V for a summary of M.F.D. equipment inventory as of 1972, including real estate.

FIRE FIGHTING WAGE RATES - 1972

Position	Rate
Laborer I	$1.80
Cookee	1.80
Equipment Helper	1.80
Radio-Tel. Operator	1.80
Laborer II	1.95
Cook	1.95
Truck Operator	1.95
Pump Operator	1.95
Straw Boss	2.10
Skidder & Other (Equipment Operator)	2.35
Timekeeper	2.35
Scout	2.35
Crew Boss	2.50
Sector Specialist	2.50
Dozer (D-6 or larger) operator	2.65
Mechanic	2.65
Sector Boss	2.75

INSTRUCTIONS
1. All positions paid straight time only.
2. No deductions for food or lodging if and when provided. Other situations will be decision of Fire Boss.
3. Cost of smoking materials and other personal items will be deducted.
4. Company personnel will be registered by M.F.D. timekeepers for hours worked on fire:
    (a) Paid by M.F.D. at base rates shown here and any fringe benefits to be paid by company: or
    (b) Paid by company in full on basis of hours kept by M.F.D. timekeeper and billing to State only on rates shown above.
5. Only those men whose names appear on M.F.D. payroll are eligible for workmen's compensation.
6. To assure payment the individual worker is responsible to report his time directly or through his crew boss to timekeeper.

Accepted February 17, 1972

Today woods crews rarely remain in camps but commute from job to home. This has been brought about by cars, roads, and other changes in working conditions. In 1937, another source of manpower in the form of the CCC camps also ceased to exist. Fire fighters now have to be recruited from Canada or locally from within the state and from the few small woods camps that still remain.

In searching through the records, one notes that over the years an astonishing fluctuation has occurred in wages paid to forest fire fighters. All of these fluctuations have an upward trend. In 1891, at the time of the creation of a forestry commission within the state, one finds the first reference to an established rate for fire fighting. It was set as "not exceeding fifteen cents per hour." It was a statutory provision, which continued up to 1908. In 1909 the forest commissioner was authorized to pay fifteen cents for each hour of service, with the addition of provided "subsistence." That rate continued through 1918, with ten cents per hour being added for board. On a ten-hour day this amounts to a total of three dollars per day, as indicated in Commissioner Colby's letter below, dated 6/1/1918, and addressed to John E. Mitchell, chief warden of Patten:

> In regard to wages of men fighting fires, let us hope we wont have any, the Law allows us to pay 20¢ per hour and each man also to be paid for his sustenance. Whenever a man boards himself of course he should have pay in addition to his 20¢ per hour for his labor and it seems about fair to allow him at the rate of 10¢ per hour for board, or in other words, where a man fights fire and boards himself for ten hours he is to have $3.00. If his board is paid by the department, no matter what it may cost, he is to receive 20¢ per hour for his actual labor.
>
> We realize the high wages that men are receiving for woods work and other labor but we cannot pay more than the law allows. We do think it is right for the Chief Warden to be liberal with the men as far as their time goes in getting them to and from any fires that may occur.

The following letter indicates that there was a problem in collecting the wages that had been earned.

Dear Sur
In regards of any labor fiting fire I haven not got any pay yet that was in June—I settled with the K.P. Lumber Co.—but they charge me my board.
An I haven got my pay for the time I fitting fire now
Pleas try an send me that money has i needs it badley
am oblige
You Struely
Answer soon

In 1921 the rate was increased to thirty cents per hour, a rate greater than the then current wages paid in the woods. This latter fact gave rise to the suspicion that fires might be set to "provide profitable employment." Such suspicions or fears were to recur often under conditions of general unemployment.

A pile of pulpwood consumed by the fire

It was not until the establishment of the M.F.D. advisory committee in 1948, that a definite policy was established for fire wages and for board. Previous to this, it was the common practice to adopt the hourly wage rate received by employees of Organized Towns and by the State Highway Commission for road construction.

I can recall meetings of the advisory committee when wage rates were adopted to make increases from fifty cents to eighty cents and from eighty cents to one dollar, representing five to ten per cent increments. The rate, by 1973, was $1.80 per hour, in marked contrast to the fifteen cents paid in 1891.

It is important to realize that unless the hourly rate met state and federal minimum standards there would be a problem in recruiting men to fight fires. As a result of the creation of the Northeastern Forest Fire Protection Commission, as well as the state and federal minimum rate standards, the finance subcommittee of the M.F.D.'s advisory committee and the commissioner came up with a realistic approach for establishing a base rate to be paid by the District, with varying rates according to one's position in the forest fire suppression organization. This system was adopted and went into effect in 1960. The new schedule was printed on cards in quantities for distribution. In this manner there was no question of the rates to be paid.

Pay adjustments had to be made in a number of instances despite this schedule. Where company personnel or woods crews were receiving more on the job than the base pay for fire fighting, the follow-

Above: Supervisor Rex Gilpatrick with jeep full of equipment especially adapted to back country fire protection
Below: Old model state vehicle used for road patrol by former Chief Warden John Mitchell, Patten, Maine 1929–35

ing arrangement was agreed upon by the M.F.D. advisory committee. Either the fire fighters were paid by the M.F.D. according to the established base rate and the difference was paid by the company by which he was employed, or the company would handle the entire payroll, taking the hours worked from the M.F.D.'s time sheets as kept by the fire boss, and then billing the District for reimbursement according to the M.F.D.'s base rate. In order to avoid delays in payrolls and especially in fighting fires of any duration, the latter procedure was usually carried out. In this way there were no hardships on the families, while problems arising from workmen's compensation and other personnel factors of employment were circumvented.

In the case of fire fighters recruited from outside the District, they became, in a sense, state employees and were paid by the District according to the base pay rate. They also received workmen's compensation in case of injury or fatality. There have been several cases of injuries where hospital and medical bills have been paid and settlement on a lump sum or extended weekly compensation has been made. The M.F.D. has been fortunate in having no serious injuries sustained by fire fighters.

Equipment rates based upon a ten-hour day were also established by the M.F.D. The rate schedule, based upon a classification of tractor, truck, car, power saw, and trailer was prepared, mimeographed, and distributed. In compiling this schedule the rate agreed upon by which landowners and the commissioner was used and proved in most instances to be acceptable to outside contractors who made their equipment available. In cases where the rates of such contractors were higher than those stipulated in the schedule, the District also paid the difference.

It should be noted that the M.F.D., not unlike the state, was operating on a businesslike, realistic budget and not on a crisis to crisis basis. Beginning in 1955, a systematic start was made in a fixed budget item for fire suppression. A special tabulation was prepared for the first time, showing the record from 1917 to 1972 of suppression costs for labor, equipment, and supplies compared with the number of fires, acres burned, and the damage.*

The issue of suppression costs on any particular fire is a difficult matter. Annual costs fluctuate markedly. One of the most crucial decisions a fire boss has to make is when to start demobilization of crews and the release of equipment once the fire appears to be under control. Such decisions have to be based upon much more than costs. The full magnitude of such situations can be appreciated through the

---

* See Appendix V for M.F.D. suppression costs (1917–1972).

EQUIPMENT RATES - 1972

	BASED ON 10 HOUR DAY	
TRACTORS (1)	Without Operator (2)	With Operator (3)
D-2, John Deere, Oliver, etc.	$ 20.00	$ 43.00
D-4, TD 6	50.00	75.00
D-6, TD 14, HD 8 & 9	90.00	120.00
D-7, TD 18, TD 20, HD 10, 14, 15, 20	120.00	150.00
D-8	150.00	180.00
Grader Cat. 12F	90.00	120.00
WHEELED SKIDDERS		
Timber Jack or Tree Farmer	30.00	55.00
LOWBEDS (4)		
Up to 12 ton - with tractor		60.00
Twelve ton and up with tractor		85.00
TRUCKS		
Suburbans, Carry-all, Sedan delivery or 10¢ per mile	8.00	
Jeep - pickup, up to one ton	10.00	
Truck, 1½ - 2 ton	13.00	
Bombardier	15.00	
Truck, 2½ - 3 ton, power wagon	16.00	
Truck, 6 x 6	22.00	
Truck, 10 wheel	30.00	
BUS AND CAR MILEAGE		
Cars		.10 per mile
18-20 men		.25 per mile
40 men		.40 per mile
PUMPS		
1½" Pacific, Gorman-Rupp, Hale		10.00 per day
Trailer, 250-500 G.P.M.		20.00 per day
CHAINSAWS		
		3.00 per day
or for 8 hours or less at rate of		.35 per hour

INSTRUCTIONS
1. An owner leaving a machine on the fire, without specific request by the Fire Boss to do so, does not commit the MFD to pay.
2. Fuel, oil and operator compensated by MFD. All other costs of maintenance, liability for loss of tractor, etc., assumed by owner.
3. Fuel and oil may be compensated by MFD. Operator and all other maintenance and liability assumed by owner.
4. Lowbeds which are being denied other work for reasons of a specific request of the Fire Boss to remain on the fire may be paid at a flat rate of $25 for a 24-hour day.

example of just one of the major fires that occurred during the disastrous year of 1952. Over four hundred fire fighters, mostly French Canadians, were employed in suppressing the great Pierce Pond fire. While most fires are of short duration, the Pierce Pond conflagration made it necessary to keep men on patrol for seventy-six days! Under such conditions there is little wonder that suppression costs fluctuate from year to year.

While this has not been a complete account, it does give the reader some insight of how the M.F.D. met and paid suppression bills for

BONDED FRENCH CANADIAN FIRE FIGHTERS ON FIRE LINE PATROL
AND MOPUP

Above: stringing hose along the fire line for mopup; *below:* making a "watermelon roll" in picking up hose during mop-up

labor and equipment. Early emphasis was placed upon detection and actual suppression, but today the trend is toward a broader endeavor, which includes a greater effort in forest fire prevention through education.

Closely associated with any program of fire prevention and proper education of the public are the studies of causes of forest fires and their impact upon forest resources.

In my opinion very little research information is available concerning the recovery of Maine lands swept over by forest fires. Important changes in the very character of the land are disguised by the forest growth that follows a burn. Depending upon the severity of the burn, many years pass before the original forest type is restored. There is a real need for more technical studies of this nature, and many old and recent burned-over areas offer the opportunity.

This suggestion is prompted by an interesting study made in 1904 by professor Samuel N. Spring, of the University of Maine Forestry faculty, in cooperation with the Bureau of Forestry, as the U.S. Forest service was known in those days. Forest Commissioner Edgar E. Ring was responsible for engaging the services of these two agencies and entrusting them with the following objectives:

(1) The study of the nature and effects of forest fires as seen from a detailed study of three areas
(2) The control and prevention of forest fires.

The three fires, all of which had occurred in 1903, were:

(1) in the vicinity of Mt. Katahdin, 84,480 acres involving townships 4 and 5, Range 9, the Wassataquoik area of Piscataquis County. Within this area two previous fires had occurred in 1837 and in 1884.
(2) in township XXII, Hancock County, 12–13,000 acres. This area also had been previously affected by fires in 1858, 1872, and 1884.
(3) in townships "D" and "E" in the Rangeley Lakes section, 30,000 acres. The Rumford Falls and Rangeley Lakes Railroad ran through this area with a station at Bemis and was involved in transporting pulpwood to the International Paper Company's mill at Rumford Falls.

Conditions surrounding the fire that burned over the third area were of special interest. Where no natural springs or streams were near the railroad, water barrels were placed beside the right-of-way and kept filled. During the fire, a strong suspicion of sabotage arose

The proper marking and storage of equipment provide for quick loading and identification on the fire lines

Canadian Pacific Railroad patrolman, equipped with gasoline car, crossing trestle at Onawa, Maine. Bearstone Mountain is in the background. First railroad patrol in 1915

when it was discovered that holes had been shot through some of these water barrels. However, no conclusive evidence was found.*

Forest fire prevention laws pertaining to railroads date back to 1891. These were greatly strengthened with subsequent legislative amendments.

Railroad officials were cooperative in the inspection of locomotives as to defective nettings (spark arresters) ash pans, brake shoes, and protective screens on windows to "prevent the throwing of burning matches, burning cigars, burning cigarettes or parts thereof from windows of such cars."

In the 1913 commissioner's report one finds the following: "No depositing of fire, live coals or ashes upon tracks by trains going through forest lands in the Maine Forestry District. When engineers, conductors or trainmen discover that fences along the right of way on woodlands are burning or in danger from fire, they shall report the same at their next stopping place which shall be a telegraph station."

When the diesel oil locomotives replaced the old steam coal-burning type, many thought this would eliminate most fires caused by railroads. However, fires continued to be set, and the need to restore this as a separate *cause* of fires on reporting forms was recognized. It should be said that railroad companies and their national

---

* A complete record of the findings of these three investigations is contained in the Forest Commissioner's Report of 1903–04, pp. 58–112.

association have been most cooperative in working on problems of engine design and the chemistry of fuel oil. National average figures of railroad caused fires are from 5 to 7 per cent.

Another aspect of railroad fire prevention was the statute for the annual burning off or removal from right of way of all grass, brush, and other inflammable material. There were also special provisions for new railroad line construction. Effective patrols on all tracks going through wooded areas were implemented, and a program began through which section foremen and railroad chiefs in charge were commissioned as deputy fire wardens by the forest commissioner, until 1967 when, as previously shown, the matter of appointments was turned over to the railroad companies.

The following table shows the number of railroads and mileages operated by each carrier in Maine affecting areas mostly in the M.F.D.

RAILROAD MAIN TRACK MILEAGES IN MAINE

Railroad	1911 (miles)	1972 (miles)
Bangor & Aroostook R.R.	627.80	572.59
Canadian Pacific R.R.	177.28	233.70
Maine Central R.R.	764.64	781.58
Sandy River & Rangeley Lakes R.R.*	103.36	--
Total Mileage	1,673.08	1,587.87

*Narrow gauge (2 feet): discontinued in 1936

Note: A large percentage of this trackage runs through forest areas. Between 1911 and 1972, considerable miles of track were sold to other companies.
Ref.: 1911 Railroad Commissions Report, pp. 6-7
1925 Biennial Report of the Public Utilities Commission, p. 38

Railroads have been most cooperative in patrolling their lines during periods of dry forest hazardous conditions. Modes of travel were gasoline speeders, putt-putts and velocipedes.

Canadian Pacific and other railroads have made tank cars available for hauling water to fires within hose line distances.

Other studies have been made concerning the recovery of burnt lands. I recall the work done by Georgia-Pacific on the three-hundred-acre Farm Cove fire in Township Six N.D., Washington County, in preparation for aerial seeding by helicopter; the work of St. Regis on some of their burned areas; that of the Eastern Corporation on the Myra-Beddington CCC road; the ten-acre experimental control burn on Indian Township conducted by the University of Maine's School of Forestry Resources; and the research work of the Massabesic Experiment Station at Alfred, Maine, on some areas of the 1947 burn. These are not isolated cases, for undoubtedly similar studies have been carried out on burned areas within the territory of M.F.D. and other parts of the state. Results, while not always conclusive, offer at least some object lessons.

In recent years, along with the development of standard reporting forms, greater emphasis has been placed upon definitions and causes of forest fires. Several years ago Fred Holt, as deputy forest commissioner, collaborated with Wayne Banks of the U.S. Forest Service in a short study of the classification of numbers and causes of forest fires. Today, data processing through the computer system makes it possible to swiftly and effectively pull out nearly any fire statistic desired for a specific purpose.

Examples of some of the uses made of such organized data would be a special fire prevention-education drive against a specific cause found to have been common to an unusually large number of fires; provision of statistics to support requests for legislative funds; analysis of costs, acreage, and damage figures; provision of vital information for lawsuit settlement cases; the comparison of state statistics to national; resource values and risk studies, and countless others.

In keeping with the effort to improve forest fire reporting, the necessary form has gone through a series of changes in format. It has been standardized by categories and is now geared to be used in feeding data to the computer.

Distinctions between categories of "cause" and of "class responsible," for example, have often caused confusion. In the 1940 Forest Commissioner's Report, one finds the following listed under cause: trappers, clearing land, adjoining town, sportsmen, fishermen, hunters, set, smokers, icehouse fire, hedgehogs, railroad, lightning, unknown, locomotives, etc. On present forms, it is possible to make a clear distinction by checking the proper box. Another improvement is the better understanding of what is to be considered a "reportable fire." In addition better accuracy is now carried out in the field in determining acreage and damage figures.*

Another most important factor in forest fire reporting has been the completion, as far as current needs are concerned, of a task force study known as "values at risk." This was a joint cooperative effort between the U.S. Forest Service and the National Association of State Foresters to establish a set of values on an interim basis for each state for every acre of forest land to be protected against fire.

Fed into the study were the factors of risk, damage, and hazard, both tangible and intangible. The end result would be a schedule of values at risk for all forest land, involving timber, water, recreation,

---

* See Appendix V for samples of reporting forms old and new, and one example of an affidavit concerning one fire's cause.

wildlife, forage, real and personal property, and, recently added, life, health, and air quality.

Not all states have the same reservoir of statistical information to make their own per acre of forest land values at risk schedule. However, a final result of the study was a model for the determination of wildland resource value that enabled each state to work up its own schedule, using the data available and applying the formula. A few years ago a figure of two hundred and seventy dollars per acre for forest land was derived for Maine as a value to be protected against fire.

During most of the nineteenth century, no records of fires were kept by anyone. This was partly due to a lack of personnel responsible for gathering data. It was not until 1903 that the state made funds available that put patrolmen on a partial state payroll basis and into the unorganized territory. Such action provided a little more control in forest fire reporting. Since then the record of the number of fires, acres burned, and damage resulting to timber and property on a state-wide scale has been systematically kept by the forest commissioner.

The accuracy of the collected data improved consistently as the Forestry Department became more and more responsible for fire control, improving markedly with the formation of the M.F.D. in 1909 and the strengthening of fire laws for organized towns in 1913. While these records of sixty-nine years of forest fire statistics (1903–1972) may appear dry and uninteresting to some, they do have a real value in recording an important part of the history of Maine forests.

Their full significance can be seen when they are correlated with factors such as cyclic weather conditions and other events that took place during the M.F.D.'s history. The factor of weather must not be discounted, for the statistics reflect the good and bad years, when seasons were either wet or dry. Occasional years of big "blow-up" fires distort the periodic averages. For example, in the bad years of 1903, 1908, and 1911, before the District really got started on an effective protection program, 398,577 acres were burned. If we add the 211,513 acres burned over in the four consecutive bad years of 1920–24 and the 130,294 acres burned in 1934, we arrive at the astonishing fact that these eight years out of M.F.D.'s 69-year history account for over *seventy per cent* of the total Maine burn of 1,054,000 acres during the same period.

Continuing a survey of the records, the smallest acreage burned was in 1917, when the figure was one hundred and forty-seven acres, while the largest, 200,232 acres, occurred in 1903. The smallest number of fires was sixteen, recorded in 1967, and the largest number, three hundred and one, came in 1952. It is of interest to point out that during the latter year one hundred and sixty-five of the fires were caused by lightning. In 1934, when the number of acres burned over totaled 130,294, one fire covering nearly 60,000 acres originated in Quebec, where homesteaders were burning brush, and spread across the border. This fire was in the vicinity of Lac de La Frontière.

Out of the total of over 8,906 fires within the territory supervised by the M.F.D. nineteen were under one thousand acres, ranging from one hundred and forty-seven to nine hundred and eighty acres, while ten of these were under five hundred acres in size. On an average, one hundred and twenty-nine fires have occurred annually within the state, including those in the area protected by the M.F.D., with an average yearly burn of 15,275 acres and an average yearly damage amounting to $78,634.

In looking at the broad picture, the following relationship of acreage burned within the territory of the M.F.D. as compared to the state's total forested acreage is of interest. Out of a state total of 17,748,600 acres, the cumulative total area burned within the territory of the M.F.D. was 1,081,248 acres. The sixty-nine year record shows that of this total burn some of the areas were reburned in subsequent years.

Of particular interest is the fact that of the total number of fires 37.9 per cent occurred within M.F.D. territory, and that these fires represented 6.1 per cent of the total area burned over and amounted to 23 per cent of the total damage to the state.

Based on state-wide figures representing the average number of acres burned annually since 1975, Maine has an excellent record of keeping within and often well below the allowable burn factor set as the goal both by individual states and by the U.S. Forest Service. At one time this goal was one tenth of one per cent allowable burn, which for Maine would amount to about 18,000 acres. With the continued improvement in forest fire protection, this factor was reduced to about 9,000 acres.

Maine's good forest fire record as reflected in the statistics is due to numerous improvements, a number of which are listed below:

1. More funds from M.F.D. from mill tax increases, for personnel, equipment, supplies and capital improvements
2. Improved communications—telephone to radio
3. Improved fire detection—lookout towers to aircraft
4. Shift from hand-dug fire lines to use of mechanized equipment
5. The Northeastern Forest Fire Protection Compact between the six New England States and New York and the Provinces of Quebec and New Brunswick, which provided for large fire organization training

6. Better knowledge of fire behavior and better application of tactics and techniques
7. Improved public cooperation
8. More stringent fire prevention and suppression laws
9. Better methods of reporting and estimating acreage burned and damages sustained.

Unloading fire equipment to be taken into fire camp

Loading fire equipment at Cross Lake Storehouse

Parachute drops of equipment and supplies on forest fire areas are now standard procedure

IX

RADIOS AND AIRCRAFT

I had to climb mountains and ridges to see where the fire was. It was then that I conceived the idea of having an airship for observation work. I would give anything for a look at that fire from above such as I could see from an airship—a moving observation tower.*

As has been made evident, the history of the M.F.D. is an account of continual modernization and the incorporation of more technical and sophisticated equipment in an endeavor to protect the forests of Maine. In this chapter we shall be concerned with two such programs of modernization, each of which has had a major effect on the structure and efficiency of the M.F.D.; namely, the development of a radio network and the use of aircraft.

In 1946, the M.F.D. had its first official introduction to radio as a tool of communication. In that year, ten portable amplitude modulated "Link" units (model 695-B), equipped with transmitter and receiving sets, were made available. These were approved by the Federal Communications Commission and licensed to operate on a frequency of 35.94 megacycles. These sets were placed in the care of the supervisors and, in a few instances, of chief wardens. They had a limited range due to their low battery power, and good results were not obtained beyond a radius of twenty-five miles.

From such a beginning, radio grew and expanded into a modern and efficient network providing a comprehensive coverage and linking lookout towers, watchmen, and patrolmen to chief warden camps and other central offices.

---

* Letter from former Chief Warden Mitchell, of Patten—1915.

The difficulties in establishing such a radio network reflect the struggle that took place during the building of a woods telephone system in the earlier years. There were many situations that challenged the imagination and ingenuity of the radio technicians.

The first experimentations with radio equipment within the M.F.D. had occurred during the fourteen years before 1946, when radio communication was still novel and its dependability far from proven. The particulars of the first attempt to install radio equipment on a mountaintop are given in the following quotes taken from a letter written to Chief Warden Duluth Wing, of Eustis, by Kenton E. Quint, one of Maine's radio and telephone pioneers, and president of the Somerset Telephone Company:

> Yesterday while you were in the office I recalled my early experience of installing crude radio equipment for the Forestry Department back in the early 30s.
>
> The story is as follows:
>
> In 1931 or 32 I became acquainted with Bob Stubbs of Strong, then a Forestry Supervisor, and we discussed the desirability of equipping lookout stations with 2-way radio apparatus. From these conversations Jack Pierce and I built a low power (5 watt) 2-way battery powered radio telephone set using a design published in QST, the amateur radio magazine.
>
> This design used a pair of type 31 1.5 volt filament tubes in a modulated oscillator circuit in the old "Ham" 5 meter band. The modulator was a two stage audio amplifier winding up with a pair of 33 type tubes providing about 3 watts of audio.
>
> The receiver, as I recall, was a superregenerative type with the low frequency oscillator on about 20kc.
>
> The receiver used the same type low power tubes, 3 or 4 type 30 tubes, plus a type 31 as output audio to a pair of head phones.
>
> The microphones were telephone type carbon transmitter heads.
>
> The equipment was powered by 4 Eveready air cell 1.5 volt batteries in parallel for the filaments and paralleled 180 volt "B" batteries for the transmitter and 45-135 volt batteries for the receivers.
>
> The transmitters and receivers were built in light weight aluminum boxes and all power batteries were put in wood boxes with plugs.
>
> After building these terminals I went to Augusta and saw Neil Violette and received his approval for a test of these equipments between Kibby Mountain and Mt. Bigelow the next sum-

mer. We also received his assurance that if the tests were successful the Department would certainly be interested to develop a state-wide system using battery powered terminals on several mountains and a number of automobile sets in supervisors cars.

In due course in the early spring of 1932—or 33, Pierce and I installed a set on Mount Bigelow where Herbert Blackwell, now retired and living in Stratton, was the watchman.

We installed a vertical antenna made of copper wire outside the window of the Mt. Bigelow tower and as I recall, we used ordinary silk covered lamp cord as the feed line between the transmitter and the antenna. God only knows what the loss was but there was still enough power to light up a 1 watt neon bulb on the end of the antenna when held in the hand.

After installing the terminal on Bigelow and hooking up a battery power broadcast radio for Herbert, we waited for good weather for the trip to Kibby.

We drove to Jackman and took the C.P. Railroad to Skinner where we stayed overnight at a set of sporting camps, now gone.

These camps were run by a couple who now own a motel in Moose River. These camps were quite nice, and it was there in Skinner forty miles in the woods where civilization really blossomed. They had lavender colored toilet paper, the first I'd ever seen.

To backtrack a little, I had cut one foot on a sharp rock while swimming in the Carrabassett River at East New Portland and was still limping on that foot, and about half way into the foot of Kibby from Skinner that cut opened up and we nearly gave up the trip. But being young, I wouldn't use any sense but kept on.

Reed Sawyer, the watchman on Kibby, met us at the foot of the mountain with another man. I don't recall his name but they relieved me of my pack and load, and I hobbled up that damn mountain whose trail had been cut clear about 30 feet wide with no small trees to help the ascent.

The others were well ahead of me and by the time I arrived they had the equipment all set up in the tower.

We then ate lunch and then called Herbert Blackwell on Bigelow thru the wire line network thru several switches at Skinner—The Chimes—and possibly King and Bartlett to Stratton where Mrs. Lee called Herbert on the mountain.

We arranged a schedule and at the appointed time I grabbed the microphone and called in a loud voice "HELLOOOOOOOO Mt. Bigelow" and wonder of wonders Herbert came right back loud and clear.

And that, I believe, was the first radio contact ever made by the Forestry Department.

That equipment stayed on those mountains over the winter, and with new air cell filament batteries performed without failure the second season; although the plate batteries were pretty well gone by the end of the second summer.

How did it all come out? Well, Quint and Pierce did not reap the harvest we had sowed with considerable sweat and some hundreds of dollars of money.

The Department was delighted with the experiment, and an agreement was made to equip 12 mountains with similar equipment. Radio licenses were applied for and materials were ordered to build more permanent terminals.

Then the axe fell. Other people wanted in the act, RCA and others.

To make a long story short, when Mr. Violette went to the governor's council to get release of the money to buy the radios, one of the councilors had a favorite in Skowhegan who wanted to supply the system. Unless he got the job the council would not approve the funds, and that was that.

Mr. Violette being a man of his word would not purchase radio gear, unless it came from Quint and Pierce, so the entire project was dropped and it was many years later after Mr. Violette was no longer Commissioner before anything was done about radios for the Department.

What happened to Quint and Pierce? Well one day while on Bigelow Pierce was talking with a "Ham" on Cadillac Mountain and the man he was talking to turned out to be an official of General Radio Company in Cambridge, Mass. From that contact Pierce went to work for General Radio, then became a teacher at Cruft Laboratory at Harvard. He was sent to Siberia on an eclipse expedition and became famous in scientific circles. Quint continued to operate Somerset Telephone Co. which he bought in 1929.

Somerset Telephone is up to its ears in radio and microwave having a major microwave repeater station on the summit of Sugarloaf Mountain and fire radio telephone systems serving over a hundred fixed and mobile dial telephones in its service area, which includes the Sugarloaf-Kibby area.

That, in a nut shell, is a brief history of the first radio for the Forestry Department about 40 years ago.

Returning to the "official" introduction and development of a radio network servicing the communication needs of the M.F.D., following the experiment of 1946 with Link equipment, the Forestry Department started the serious task of developing an efficient and well-organized system of radio communications.

Initially the Department, with the authorization of the governor and his council, placed orders for sets built by Motorola. This was a logical move, as the state police were using this same equipment, a fact which allowed the two agencies to exchange parts, share experiences, and utilize each other's test equipment. Another saving to the Forestry Department was the pooling of orders for radio equipment with the U.S. Forest Service. Later this practice was discontinued because of difficulty in obtaining the specific equipment required by the M.F.D. As competition developed within the field of electronic manufacturers, the Forestry Department purchased equipment made to specifications from several other companies besides Motorola, including Comco, Dumond, and Radio Specialty.

One major problem in establishing a radio communication system within the remote areas of the District was the lack of electrical power. It was here that the ingenuity of the radio technicians was truly tested. To meet this need, a composite battery was designed by the Burgess Company. The first type of battery developed had a five-hundred hour capacity and weighed eighty pounds. Such an electrical supply could last a season. The problem was how to back-pack these batteries into the remote places where they were needed. Later, a three-hundred hour capacity battery weighing some twenty-eight pounds was designed by Burgess, but these batteries had to be renewed three times to last a season. All such batteries were made in quantity and stored in a refrigeration plant in Waldoboro rented for that purpose.

Before the advent of these batteries various methods of recharging electrical cells were tried, including motor and wind-driven generators, but they proved both expensive and cumbersome. In the earliest stages, the source of power was dry cell batteries connected in series, a source which was not only cumbersome but costly.

With the establishment of radio communication, a new era of efficiency and safety came to the lonely posts of lookout towers. An ingenious remote control system was established between the towers and the camps of the watchmen.

When a watchman left his tower at night to go down the mountain to his camp, there was always the chance of accident along the trail. With the remote control equipment, the watchman could turn off his radio on leaving his tower and on reaching his camp flip a switch that allowed him to contact his chief warden. Such equipment was not only a safety factor, but resulted in a saving to the District of thousands of dollars.

As radio became an established part of the M.F.D.'s communication system, attention was focused on ground to aircraft communication. Initially it was a matter of placing self-contained, battery-operated equipment in the plane. Later, twelve-watt mobile units were installed and connected with the battery system of the Cessna planes. Still later, thirty-watt units transmitting on the high band frequencies of two channels and then four channels were employed. As part of a cooperative plan, planes of the Fish and Game Service were equipped during periods of emergency with radios operating on the same frequencies and powered by dry cell batteries.

A further expansion of cooperation within the radio network included hook-ups with radio equipment within Baxter State Park, Allagash Wilderness Waterway, certain other parks of the State Park and Recreation Department, and on the basis of limited use, with the Maine National High Adventure Program of the Boy Scouts of America located at Matagamon and at Seboomook.

An important system of radio liaison existed in the private sector between the various lookout towers and the radios of the pulp and paper companies. Such a system was eliminated, of course, with the phasing out of the towers. Companies which cooperated in this venture were:

- (a) Penobscot Development Company, in cooperation with the towers at Medford and Mount Chase
- (b) Eastern Pulpwood Company, with Musquash Mountain
- (c) Georgia-Pacific Corporation, with Cooper Tower
- (d) St. Regis Paper Company, with Wesley District Headquarters
- (e) Great Northern Paper Company, with Squaw and Spencer mountains
- (f) Brown Company at Cupsuptic Headquarters for tie-in with operating camps in Cupsuptic and Parmachenee and the Berlin, New Hampshire, office
- (g) International Paper Company, with Priestly Mountain, and on into Clayton Lake.

Such connections were extremely important in maintaining close contact between landowners and the M.F.D. for purposes of coordinating manpower and equipment during on-going fires. They were also the means of reporting the daily weather and forest fire class danger days.

In 1958, the F.C.C. announced a new regulation that had a direct effect on the growing radio network within the M.F.D. Previously, the Forestry Department as well as the District had operated on a state-wide, low frequency band, using 31.620 and 31.740 Mega Hurtz. The new F.C.C. regulation demanded that all the Department's radios be narrow band. This posed a serious matter of readjustment.

After a series of meetings with Mr. Claypool, radio technician of the U.S. Forest Service's Radio Laboratories at Beltsville, Maryland, a decision was reached. Only forty per cent of the Department's existing equipment could be converted to narrow band, leaving sixty per cent that must be replaced. Moreover, the low band frequencies within the state were already overcrowded, a condition that hampered dependable and effective transmission. There being no more low band frequencies available, it seemed best to apply through the F.C.C. for a high band allocation of five channels—159.33, 159.36, 159.39, 159.42, and 159.45 Mega Hurtz (M.H.I.). Such an allocation would allow four area frequencies and one state-wide frequency to be used for administrative purposes. The application made to the F.C.C. was granted and the Department was also permitted to retain the two low band frequencies for administrative purposes.

A deadline of 1963 had been set for the conversion, which was to cost the M.F.D. $115,000 and the Organized Towns another fifty thousand dollars. It was, however, a wise move, for the high band offered protection from interference for the next ten years at least. There are now rumors that the allocated frequencies to the forestry and conservation agencies may again be encroached upon. Already the battle is joined between these conservation agencies along with other groups and the F.C.C.

To service the present District radio system properly takes considerable personnel. There are three radio technicians headed by a director of communications. These technicians are available for use in the Organized Towns as well as in the M.F.D. The laboratory shops are presently located at Bolten Hill headquarters, Augusta, with a branch at Island Falls. Such facilities represent a marked improvement over the first repair shop, which was opened at Windsor, Maine, in 1947.

Over the years the radio communication network became a major part of the M.F.D.'s program of modern forest protection. At one time it was the suggestion of Governor Curtis that such services be provided by New England Telephone and Telegraph on a contractual basis. This matter was considered and a figure of thirty-five dollars per unit was offered. Though similar agreements had been made in other states, it was found not to be a feasible plan in Maine because of the remoteness of many areas within the M.F.D. In addition, there was a serious question of whether services could be maintained at irregular hours and during holidays and weekends. It was the consensus of the Forestry Department as well as those of Highway, State Police and

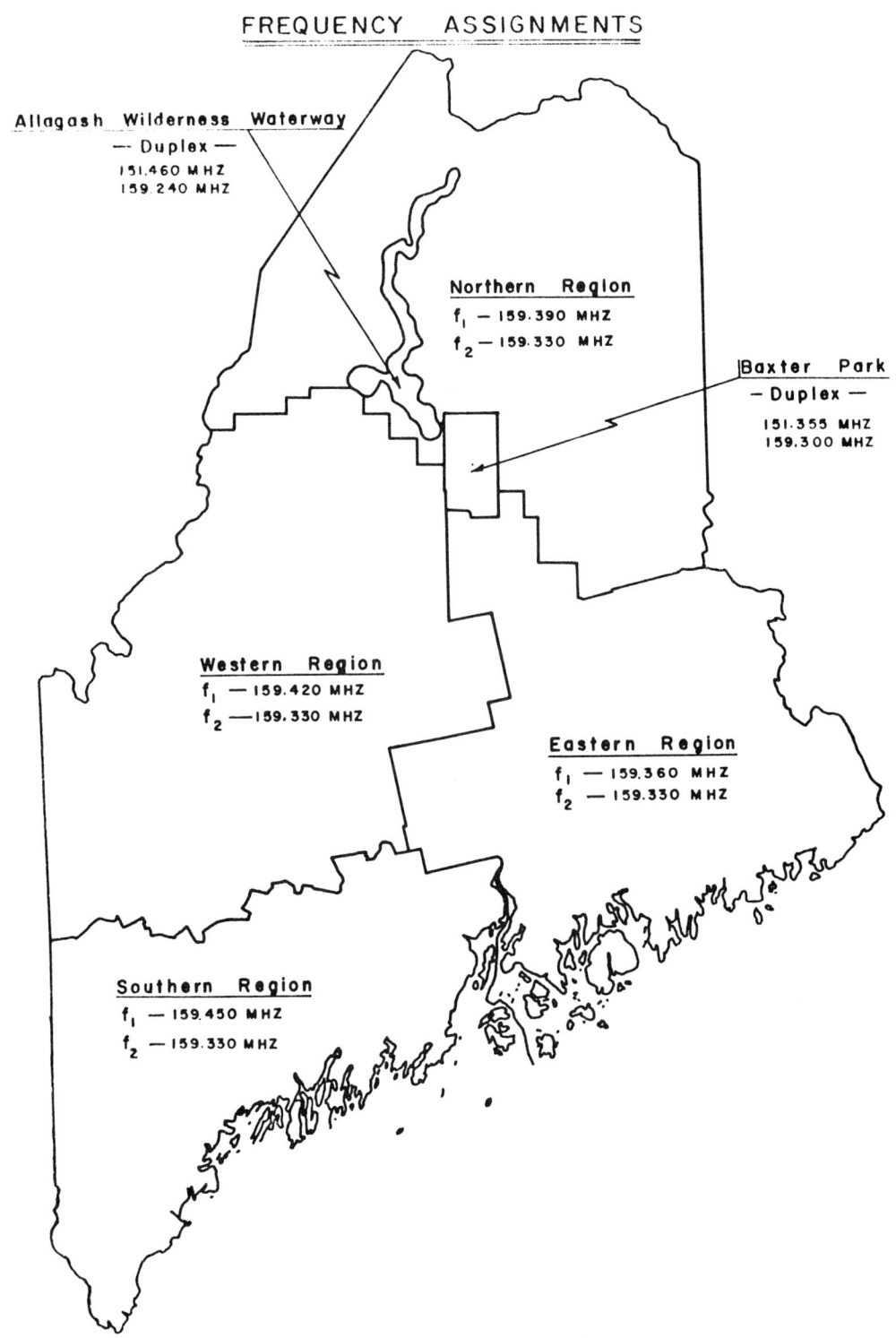

the Civil Defense that they should continue their present system of operation.

A systematic plan of radio call numbers for base stations, towers, airplanes, trucks, and cars has been in effect for a number of years. Lists of such call numbers are printed and updated annually in the *Directory and Radio Manual*. In each of the four regions of the state, blocks of numbers have been assigned. The one hundred series goes to the eastern region, the two hundred to the western, the three hundred to the northern, and the four hundred to the southern region. The numbers one and two are reserved for the forest commissioner and his deputy.

The radio call numbers have been incorporated into the vehicle licensing system. The number appearing on the plate serves as vehicle identification as well as radio call. Call numbers painted on all trucks and cars have been especially helpful during fires as an aid in ground to air communications, being plainly visible from aircraft flying over the fire areas.*

The present radio system in the District represents a big investment. The following schedule compiled in 1972 for insurance purposes is of interest:

RADIO INVENTORY AND VALUE 1972

		Number	Dollar Value
A.	M.F.D.		
	Handie Talkies	100	$ 60,000
	Land Mobiles	125	137,500
	Mobile units	166	99,000
	A. C. Base Stations	55	82,500
		446	$379,000[1]
B.	Radios Used in Organized Towns	175[2]	164,000
		621	$543,000
C.	Capital Equipment Value		11,681[3]
	Total		$554,681

[1] Represents 70 per cent of total Forestry Department investment in radios.
[2] All types of units grouped together.
[3] Capital equipment: antennae, test equipment, etc.

NOTE: These are updated figures from original schedules sent to the State Insurance Department and were requested by the Governor for the recent State Cost and Management Study.

* See Appendix VI for M.F.D. schedules of radio call and car plate numbers for 1953 and 1972.

While this story of the conversion from woods telephone lines to radio communications has been lengthy, it should be interesting to the old-timers who can remember this change.

Just as in the case of the introduction of radio communications, the appearance of aircraft produced a major change in the M.F.D.'s program and capability. And just as in the case of radio communication, experimentation with the use of aircraft took place over a number of years.

The first idea for the use of aircraft in forest fire protection in Maine, which might be considered a "first" in the entire country, is found among some old letters written in 1915 by former Chief Warden John Mitchell of Patten, Maine. It is to be remembered that only twelve years had elapsed since Orville Wright had made his first flight at Kitty Hawk.

In one such letter Mitchell wrote: "I had to climb mountains and ridges to see where the fire was. It was then that I conceived the idea of having an airship for observation work. I would give anything for a look at that fire from above such as I could see from an airship— a moving observation tower."

In 1916, Mitchell got his first airplane ride over some of his district. It made a tremendous impression upon him, and in succeeding years he became very persistent in his efforts to promote the use of airplanes in the M.F.D. He wrote many letters, held interviews, got recognition in a feature article published in the *Boston Herald* (January 30, 1921), appeared before the Maine Legislature in defense of an unsuccessful resolve to raise $25,000 for the purpose of introducing air service into the forest protection program, and kept contacting Forest Commissioners Samuel T. Dana and Neil L. Violette, occasionally receiving a mild note of censor from the Augusta office for his overzealousness. The following letters, written by Fred A. Gilbert, of Great Northern Paper Company, in December 1922, and Forrest H. Colby, lumberman from Bingham, on New Year's Day, 1923, exhibit some of the cool reception that Mitchell faced. He was just too far ahead of his time.

> Acknowledging yours of December 17th regarding use of Airplanes for forest protection. Would expect in due time they will come into substantial use. It may be a little early to consider them; as the value of forest products go up the more we can afford to spend to protect them. At the present time with taxes as they are and everything accordingly high I expect to find citizens very careful about taking on new obligations, and it ought to be so. We are not anywhere near normal yet, and would think it better to plod along the old way until we had a good

substantial bank account that would enable us to try the new. This is about the way people handle their own personal matters.

With kindest regards and best wishes for a happy and prosperous New Year. . . .

Your letter of December 23 was received in due time. I am sorry for the delay in answering, but you know how busy it is at Christmas time and I have been away from the office quite a good deal of the time.

I know you have always been very anxious for the Forest Service to have an aeroplane and I presume probably you are right but you asked for my candid opinion and up to the present time, I cannot agree with you.

We had a good trip to Boston and the Forestry Meetings were first-rate, although, as usual, there was a whole lot said that I did not agree with.

I shall hope to see you at Augusta or somewhere down this way some time this winter.

Wishing you and yours a very Happy New Year. . . .

But while there was reticence on the part of many foresters in the State of Maine, Mitchell was well aware that the use of aircraft was progressing in other parts of the nation. He heard from the U.S. Air Service headquarters in Boston, receiving ideas and suggestions. Reports came from work being done in California and in Oregon. In 1920, Mitchell learned that the Brown Company of Berlin, New Hampshire, had two "aeromarine" planes for sale, which had been used for cruising timber in Quebec. With such encouragement and informatioon, Mitchell persisted.

His most staunch supporter was George W. Maxim, of Winslow, Maine, a local commercial pilot who did considerable promotion work. Eventually their efforts began to pay off.

Maxim (later associated with the Curtiss Flying Service, Inc., of Garden City, Long Island, New York) appeared before groups of landowners and wardens in both Augusta and Bangor during the spring of 1927.

In that same year, Forest Commissioner Violette, being concerned about getting the necessary funds for introducing aircraft on a trial basis, appointed the following committee to look into this matter:

Ames, Alfred, of Machias, Maine
Braman, R. A., of Portland, Maine
Colby, Forrest, of Bingham
French, Jerome, of Eastern Manufacturing Company

Lannigan, William J., of Hollingsworth and Whitney
Lockyer, Scott, of Brown Company
Lovin, Roy, of Calais
Mullaney, R. E., of Eastern Manufacturing Company
Pierce, James, of Houlton, Maine

The recommendation of this committee was for the M.F.D. to go ahead with a contract with the Curtiss Flying Service for a seaplane and pilot at sixty dollars an hour for one hundred hours of actual flying time. The plane was to be hired as an experiment in forest fire protection. On May 13, 1927, Supervisor George Gruhn, as official observer for the M.F.D., and pilot George Maxim flew to New York and brought the contracted seaplane to Greenville by the latter part of the month. The plane was used for approximately eighty hours, with most of that time consumed in flying chief wardens over their respective districts, but actual fire work was also done on three fires and particularly on the big Chase Stream—Moxie Gore—Indian Stream—Squaretown fire, which covered some six thousand acres and occurred on the twenty-third of June.

Chief Warden John Mitchell got his first trip in this plane on June first, flying out of Shin Pond. Forest Commissioner Neil Violette wrote John the following letter after this flight on June 8, 1927:

> I'd like to have you write me a full account of your trip in the plane, whether or not you consider it was worth while, and that the plane will be beneficial in this work. Could you, in your short trip, pick out your lookout stations, trails, camps and other spots?
>
> Anything you care to tell me about your trip or any suggestions that you may care to make, will be greatly appreciated. As you know, the plane is merely an experiment and it is up to you chief wardens to determine whether or not you think it is worth while.

John Mitchell's reply was dated June 19:

> Dear Neil:-
>
> I am writing to you to express my opinion of the use of the Hydro-plane for forest protection.
>
> Now Neil you know I have been looking for eight long years to see a plane flying over the Forestry District. And on looking up and seeing the plane flying toward Shin Pond, on May 31st, I felt that one of my greatest ambitions had been realized, and on grasping the hand of the pilot George W. Maxim, who has helped in many ways to make my dream a reality, I said George

she has come at last, and this is one of the happiest days of my life.

On June 1st with George W. Maxim as pilot, I flew over the East Branch District over which I have had supervision for thirteen years. Starting from Lower Shin Pond, on the trip around over my territory, altho it was cloudy we were flying below the clouds, the Plane is far better than a lookout, because it is a moving observation tower. During my flight I could readily pick out the different Lakes, Ponds, Rivers and Brooks. I could also discern the different vegetation, that is the Spruce Hemlock and fir. I could not tell one from the other at that height, but could tell it was soft-wood growth. The Valleys, Ridges, and Mountains can be seen as good as from a stationary lookout. Also could tell the cedar swamps, and the hardwood growth on the ridges could easily be seen. On looking down I could see the Forestry camps and see the roads, especially the Sebois stage road running from Patten to Trout Brook Farm.

Altho there was no smoke in my territory I could have very quickly seen one, and been able to tell what kind of growth, and just where it was located, if water was available for immediate use, and the shortest route to this possible fire. The day I was flying it was quite rough, but having great confidence in the pilot put my whole time looking over the territory.

Now Neil as has been already demonstrated the plane is a great asset to forest protection work, and will enable the fires to be gotten at in their earlier stages, and find what would be the best way to extinguish the fire if by water, with the pumps now available, which you have had placed in several places in the forestry district, for the use of the various Chief Wardens, or other means of fighting it. All plans for fighting the fire could be made from the plane. And to think of the difference in time saved by using the plane, and trusting to the old method of locating a fire.

And Neil much credit is due you, for we know you have been heartily in accord with this project, and have spent much time and thought in considering the plane proposition both from a useful and financial standpoint. Don't you think it has already been shown? that the experiment will be successful, and show to the land owners and citizens of the State of Maine that the plane has become a part of our Forest Protective System, which we aim to make the greatest Protective System in the world.

With the experience of the pilot Mr. Maxim has had I think you were very fortunate in getting him for this work. And Mr.

Float plane picking up water for drops on a fire

Water bombing forest fires by Forestry Department aircraft is an effective tool in suppression work

> Gruhn as observer is right on the job and sure understands his business.
>
> I note by the clipping sent to me that Chief Warden Hilton has already had some experiences in locating a fire with the plane. I am glad to note that it is your idea to get the Chief Wardens into the air with the greatest speed, so they can lay their plans at once and get the exact locality. I really think that in the near future we will be fighting our fires from the planes with chemicals.

Mitchell's concluding sentence was nothing less than a prophecy that would come true in many states.

Two incidents occurred during that first season of experimental air service in the M.F.D., that belong in this account.

On June 19, 1927, the plane was working on a fire near the south end of Caucomgomuc Lake. Later in the afternoon, the plane took off in squally weather and had risen about 150 feet when it dove into Moosehead Lake near Ledge Island. It apparently hit an air pocket. Pilot Maxim, Supervisor George Gruhn and Chief Warden Frank Conley were picked up by boat and suffered only minor bruises. The damaged plane was towed to Greenville. From this experience, it was decided to get a larger and more powerful seaplane.

On September 5, 1927, a most unfortunate double tragedy occurred. Pilot Maxim and Amos Thibodeau, Jr., of Greenville, flew to Caucomgomuc Lake to pick up Stephen Wheatland who wanted to fly over some of his land. The rest of the story continues as quoted from the files:

> Mrs. Bridges brought Mr. Wheatland out from his camp on the lake shore to the plane and the three men proceeded for take-off. Only Mr. Wheatland survived. The plane was turned into the wind on the lake and got under full speed for take-off on a windy day. The pontoons either struck a ledge or a large wave sufficient to break one of the struts supporting the plane. The engine was stopped and the plane began at once to settle in the water. They were about one mile from shore and had only a small life cushion—life preservers and an air boat had been left at Greenville. Maxim couldn't swim, Thibodeau was considered a good swimmer. All three got out on the wings for perhaps 15 minutes. Mr. Wheatland swam ashore, the other two disappeared. Annis Bridges went up the lake in his boat shortly after the accident but did not know about it, he saw neither a plane nor the men due to the waves. Neither could Mr. Wheatland see the other men or the plane or attract Bridges' attention. He got to shore,

walked to Nichol's camp on Round Pond and phoned the Bridges at his camp who went and got him.

The vision of Chief Warden John Mitchell in 1915 first saw reality in 1927 and was from that date to continue to materialize. Advances made since then in aerial forest fire protection have been phenomenal, in respect to the design of aircraft, engine horsepower, pontoons, tanks for water dropping as well as chemicals, the transportation of both men and equipment, and, finally, the use of parachute and "free-fall" techniques.

The different types of planes owned by the M.F.D. is well documented. After the Stinson plane contracted with the Curtiss Flying Company in 1927, the M.F.D. purchased its first aircraft in 1933. This was a Ryan. It was followed by a series of trade-ins and new planes from such companies as Ryan, Stinson, Beechcraft, Piper Cub, Taylorcraft, Seabee, Luscombe, Aeronca, and Cessna. To this list must be added the federal excess Beavers and Bell Helicopters. Today the M.F.D. has seaplane bases at Greenville, Portage, and Old Town, the latter having the facility for repairs, overhauling engines, and for winter storage. All of which would be both fascinating and gratifying to such pioneers in the use of aircraft as Mitchell and Maxim.

The M.F.D. did not renew the contract with Curtiss after 1927. One reason was that the season of 1928 proved very favorable in relation to forest fire conditions. In 1929–30, arrangements were made with Consolidated Airway for limited charter service for flights over Penobscot Watershed, Rangeley, and Sebago Lake regions. There were also agreements with several private companies for support planes should such be needed.

The whole history of M.F.D.-owned aircraft is centered on the exclusive use of float planes. The unorganized territory under the protection of the District was geographically ideal for pontoon jobs with its characteristic of an almost perfect network of connecting waterways consisting of lakes, ponds, rivers, flowages, and deadwaters. Such features provided areas for landings and takeoffs that proved to be extremely valuable in forest fire protection.

Something should be said of the pilots who flew the planes of the District. Much credit goes to Earl Crabb, who for many years flew the early models operating out of the Cobbosseeconte Lake base in Augusta. His pioneer work of servicing single-handedly the many requests of the supervisors was no small accomplishment. Pilots Charlie Coe, George Johnson, Charlie Robinson, and Glenn Sherman also deserve special mention. They were not only capable flyers of pontoon planes, but were skilled in handling the Beaver, Cessna, and helicopter, especially when carrying out the hazardous parachute drops

One of the Stinson planes in Spencer Bay, Moosehead Lake

and water dropping operations on fires. A most enviable record of hundreds of safe flying hours have been logged by these men over the years.

Charlie Coe was interviewed concerning his work in the early days:

When I went to work for Forestry most of my job was transporting men to maintain the miles and miles of telephone line they had. They could do more work in a week that way than they could do in a month if they had to travel by boat.

I also came to town (Greenville) and got supplies for them and for the men in the lookout towers. Those men were lucky to get out once during the summer. They stayed right there until October. I'd fly in, get a list of groceries from the tower, fly to town, get them, take them to the watchman, get a list for next week.

In dry weather we made big circles over the woodlands and then they had quite a network of towers.

After they got airplanes they could fly men and equipment into the nearest pond or river and sometimes brook nearest a fire. They'd go off to the fire and you'd fly around the fire, tell them what it looked like. We had no way to put any water on it then.

So I flew gas and pumps and hose and Indian pumps and food and men and then acted as a scout.

Parachute drops of supplies to watchmen on remote towers

We had one fire up in Poland Pond just this side of Allagash Lake.* I started moving people into it. Dick Folsom was up there with an airplane and a guy from Portage with another airplane, five airplanes operating out of the pond, and we hauled 90 men, 10 gasoline powered pumps and their fuel, thousands of feet of hose, 50 or 60 Indian tanks, shovels, fire axes, groceries, two canoes, two outboard motors, all moved in by air. And we moved it within two days. I had a 90 Champ, Dick had an Aeronca Sedan, a fellow had a SeaBee and I don't know how he got in and out of there but he did. That's an amphibian with a pusher engine built by Republic, a big flying boat hull.

I was based at the Tramway, just about 15 minutes from Allagash Lake. We had fires up on the St. John and another up near Allagash at the same time. There also was a fire on Campbell Brook and that was one of the closest experiences I ever had.

There was a huge volume of smoke and a lot of flames and they formed two fingers. I flew right down between them, fairly high, to look and see what the middle of the fire looked like. I flew along through the fires and just as I looked, a volume of super heated air rushed up and exploded into a great big ball of fire.

The flames started about 500 feet below me and went up to 600 feet above me. It was three or four seconds behind me. My aircraft was fabric and I would have been incinerated instantly. I didn't go through there again.

Fighting forest fires still is a big part of a bush pilot's business now. But now you really fight it, water bomb it, where before you just scouted and took the men and machinery in.

It was inevitable that the time would come when the M.F.D.'s lookout towers would be phased out and replaced by a systematic plan of continual, commercial aircraft services for forest fire detection and surveillance. The change-over can be attributed largely to the difficulty of recruiting people for lookout tower jobs and to various economic factors.

Finding watchmen began to become a problem in the early 1960s. Experienced woodsmen customarily employed for such positions were no longer available, and members of the younger generation were not particularly interested. At one time the romance and glamor of a summer watchman's job, whether real or fancied, were attractions, but such inducements could not compete with the increased interest in working hours, pay rates, fringe benefits, and living

---

* Coe worked on the Poland Pond fire in 1952—the year of so many lightning fires.

conditions. The watchman's job ceased to be a stepping stone in the warden service leading to a forestry career. Thus it became increasingly difficult to employ people in a job which called for long irregular hours of tower duty and for self-sufficiency, which involved housekeeping, cutting of fuel wood, as well as work on telephone lines.

Recognizing the problems, the advisory committee of the M.F.D. approved in 1967 the recommendation of the forest commissioner that contractual agreements with commercial plane owners be made on a trial basis for an aerial detection program that eventually would replace the lookout tower system. The first venture in this new program commenced in a wide area, including the eastern region of southern Penobscot and Aroostook counties and almost all of Washington County. The contract for surveillance was with a local plane service at Old Town.

Quite understandably, some of the landowners were hesitant to change after so many years of lookout tower service that had proved so successful in guarding their forest holdings. However, their concern was soon alleviated as the air patrols proved their effectiveness. There was no loss of efficiency in fire detection.

With the success of this trial use of aircraft for wide-range fire detection in 1967, the conversion plan spread to other areas of the M.F.D. From the records in Augusta and field offices, the following schedule has been drawn, showing the reduction of lookouts and the increase of aircraft flights.

Year	Active Towers	Tower Service Discontinued	No. of Plane Flights
1966	63	-	-
1967	59	4	1
1968	55	16	5
1969	39	13	7
1970	26	7	8
1971	19	3	10
1972	16	3	11*
1973	13	-	11*

Total: 46 replaced

*One flight patrol was handled by the M.F.D. supercub plane in the Northern Region.

Note: In 1968 the five aircraft patrols with 1,916 flight hours cost $28,759 against the approximate cost of $53,313 for the 20 towers discontinued in 1967 and 1968 - a saving to the M.F.D. of $24,554. By 1969 the total net saving over lookout tower staffing was $71,964.

A word here on the private commercial plane air patrol contracts: Starting in 1967, the contract was a new experience for both the M.F.D. and the flying service company. A minimum of three hundred flying hours was guaranteed at thirteen dollars per hour, with another rate for any flying time over that figure.

As both the District and the flying services gained experience, the procedures for awarding the contracts became more formal. Today bids are sent to eligible names on the State Purchaser's register, with specifications for aerial detection of forest fires. Contracts are awarded after competitive bids are carefully reviewed. In the beginning, contracts were awarded on an annual basis, but later this was changed to three-year contracts, which proved to be an added incentive for eliciting bids.

The contracts carried some very specific conditions. It should be remembered that though these were state contracts, the reference below is to the special form issued on behalf of the M.F.D.

1) Guarantee of 300 hours of flying service at a fixed rate (currently $20) between April 15–November 40 and contracted rate for flying time beyond the 300-hour minimum. M.F.D. reserves the right to use all the hours up to the 300-hour minimum regardless of the nature of the services required. Preference will be given to planes with floats.
2) M.F.D. shall be held free of any liability to the pilot or aircraft in case of accident.
3) Flight patterns and number each day between 10 A.M to 6 P.M. shall be based upon the *forest fire class danger day* as determined by the Regional Director. Flights may be made for other than detection purposes when called by the Regional Director at any time of the day.
4) Observers from the M.F.D. may accompany the contractor's pilot as designated by the Regional Ranger. "The pilot will patrol for, detect and report forest fires to M.F.D. stations by radio and follow flight patterns or provide other flying services as required by the Regional Ranger."

For the first time a five-year record (1969–73) is shown in the tabulation below of private plane contractual detection hours and all purpose flying hours by Forestry Department pilots.

The overall result of the conversion from lookout tower to aircraft surveillance is very clear. Substantial savings have been made in watchman salaries, in the areas of tower and camp maintenance costs, and in communication expenditures.

Also the use of aircraft in aerial detection has proved dependable, with no loss of efficiency while providing more accurate coverage of

HOURS FLOWN BY CONTRACTUAL AIR PATROL AND FORESTRY DEPARTMENT PILOTS

WESTERN REGION

Year	Hours flown by Contract	Hours flown by M.F.D. Pilots
1969	684	267
1970	835	377
1971	935	397
1972	669	327
1973	660	325
Totals	3,783	1,693

EASTERN REGION

Year	Hours flown by Contract	Hours flown by M.F.D. Pilots
1969	548	486
1970	504	482
1971	569	463
1972	661	471
1973	601	635
Totals	2,883	2,537

NORTHERN REGION

Year	Hours flown by Contract	Special Patrol*	Hours flown by M.F.D. Pilots
1969	550	356	-
1970	754	382	348
1971	844	577	329
1972	607	408	335
1973	924	542	300
	3,679	2,265	1,312

*Special flight patrol by Forestry Department plane and pilot.

wider areas of M.F.D. territory. Several states are now one hundred per cent dependent upon aerial patrols, and in Maine a complete change has been largely affected within the M.F.D.

The current inventory shows an M.F.D. air force of six Beavers, two Cessnas, one Supercub, and five helicopters. Augmenting this strong arm of forest fire control is the excellent cooperative support provided by Inland Fisheries and Game Department planes, along with the chartered planes owned by commercial flying companies.

Pontoon plane dropping water

Private industry planes, whether owned or chartered, have made valuable contributions in aerial patrol and in assistance missions on fires. Finally, larger aircraft from the Province of Quebec, when the Compact between Canada and the northeast was invoked, were capable of dropping one thousand to one thousand five hundred gallons of water.

"Smoke jumping" has not been developed as a part of the M.F.D.'s fire fighting program. The reasons for this are many. Not only is the equipment and training necessitated by such a program expensive, but so is the large amount of "standby time" that must be involved. Another reason for not developing this line of defense in Maine lies in the topography of the Maine wilderness. With the great number of waterways, lakes and ponds, the float plane has proven most effective in rapidly reaching fire areas. While there is no question of the value of smoke jumping in some areas of the country, there is question as to this method's feasibility in the M.F.D.

Much more could be written on the use of aircraft in the M.F.D., but the salient points have been covered. Today their value has been proven, both in surveillance and in direct air attack, and in the savings enjoyed over earlier more conventional methods.

All lookout towers were closed to the public during World Wars I and II in the interest of national security

X

WAR TIMES AND THE C.C.C.

And they shall beat their swords into plowshares, and their spears into pruninghooks. (Isaiah 2:4)

Few people today are aware of the deep concern expressed by the military and the Department of Justice in Washington relating to the vulnerability of the forests of Maine against sabotage during the global conflicts of World Wars I and II. The concern was particularly keen in respect to the vast contiguous unorganized territory of over ten million acres of forest under the protection of the M.F.D.

Two existing letters written by former Forest Commissioner Forrest Colby in 1917 and in 1918 to active Chief Warden John E. Mitchell of Patten, Maine, reveal the anxiety that was so deeply felt during World War I. Though similar letters were mailed to other active chief wardens, only these two remain to give witness to the actions taken during those disturbing times.

Letter number one:

> This is a confidential letter and I trust that you will not only hold the contents of the letter in the strictest confidence, but that you will also make as little show with the articles which are being sent to you by express, prepaid, and mentioned below, as you possibly can in the performance of your official duty. In a certain part of the State it is *certain* that at least two German spies have been in the woods and while there was no mischief done by them, it is quite certain that they meant to do harm had the weather conditions been so that they could. These men were caught and properly taken care of by certain members of our

own force, with the assistance of the Sheriff and members of his force.

After careful consideration and consultation with the proper authorities it seems wise for this department to equip each one of its Active Chief Wardens with a good revolver, ammunition, and holster for the same. If you care to have a belt for your holster, you may buy such as you think best and put the cost in your next bill. I do not feel that it is necessary to go into details with you, how and when to use these articles, should the occasion ever require, but rather leave it to your own good judgment. Suffice it to say that we do not want any German spies or alien enemies to "put one over" on any of us of this department. The Forest Commissioner and his Chief Wardens, and their Deputy Wardens, are guardians of the forests of the State of Maine. See to it that they are not only protected from fire, but from enemies that may come to us from the nations at War with us.

Letter number two:

CONFIDENTIAL.

Last season we sent you a revolver to be used while you were on duty, in case you came in contact with an alien enemy who might be trying in any way to destroy the forests of our State. I want to congratulate each and every Chief Warden upon the strict confidence in which this trust was held by you all. This was an order from the Federal Department and was really being carried on as an indirect branch of the Department of Justice. At the time we furnished you with a revolver we had planned to also furnish you with a pair of handcuffs, but were not able to procure them last season.

Under separate cover, by parcel post insured, we are sending you a pair of handcuffs today to be used in connection with your revolver should occasion require. We have ordered, which we expect to receive and distribute sometime the first part of June a revolver for each Lookout Watchman in your section. These will be sent direct to the Watchmen as soon as we receive them.

The term "active" as applied to the title of chief warden in Commissioner Colby's first letter is important. Hosea B. Buck of Bangor, a chief warden-in-charge, gave the following definition, which is well worth our notice: "By active I mean these men who were on the work continuously during the season and were held responsible for the efficiency of the patrol in the special territory which was assigned to each." From this definition it is evident that the Department acted with credible restraint in issuing arms despite the alarms that were abroad.

In 1918 similar equipment was issued to sixty-one active watchmen in the various regions. Records are lost as to the exact purchase price and distribution of these items; however, they were carried in inventory for many years. In the course of time, the revolvers slowly disappeared, being either given away, sold, or condemned. Then, in 1964, a chief warden was challenged by a State Fish and Game warden for carrying a loaded revolver. While it was perfectly proper for a chief warden as a law enforcement officer to carry one, the Augusta office issued a directive recalling all remaining revolvers. There were only nine, which were sold at public bid for thirty-five dollars each. Thus ended a chapter which had begun forty-seven years before in a time when enemy agents were suspected of plotting the destruction of one of the nation's chief resources.

Still another matter during World War I is worth noting. There was a serious shortage of lumber, which was vitally needed for the construction of trenches and other military purposes. In 1917, former Forest Commissioner Forrest Colby was made chairman of a committee for purchasing equipment relating to logging outfits and for the enlisting of Maine men to support such equipment. The project was to send lumbermen and portable sawmills from New England to Old England to assist in the production of lumber. Sawmill men and lumbermen of New England were considered the best in the business.

Final results were the mobilization near Boston of three hundred and sixty men, one hundred and twenty horses and ten portable sawmill units, which were safely landed in England. The total expense of this operation was $160,000, with each New England state contributing $12,000 for a total of $72,000 and the balance provided by private subscriptions from landowners and wood-using industries within the New England areas.

The following quote gives a clear picture of the problem in logistics required in completing this contribution to the war effort.

Harold O. Cook, later Chief Forester of Massachusetts, recalls:

> Although I thought I had done about everything in my work for the State Forester, nothing in my past career prepared me for the task as the chief stevedore in charge of loading portable sawmills into cargo ships lying alongside piers in East Boston. The powers that be figured it was better to have a man on the spot who could recognize such items as brush hooks, cant dogs or logging chains, than an experienced freight loader.
>
> The ten sawmills were assembled, loaded and shipped within 30 days of the time we started on this project. During that time, I helped to load on the ships the sawmills, 360 experienced New England woodsmen, all the equipment from sawmill boilers to

pots for baked beans and last, but not least, 120 horses. A few months later Rane heard from Atwood that the sawmills and men we had sent overseas were producing wood and lumber twice as fast as the English had ever been able to turn out with comparable equipment, and that the Massachusetts Unit had set a record by producing 18,000 board feet of lumber in one day. Some of the sawmill units went to Scotland and were set up on the Carnegie Estate at Ardgay, County Rosshire.

Just after the sawmills had been shipped to England, the U.S. Forest Service requested our office to help recruit men for two forest battalions to be organized under the Corps of Army Engineers. These units were named the 10th and 20th Forest Engineers. Rane was appointed to be the recruiting agent, and we provided office space for an Army officer who had been assigned to this duty. Our office interviewed about 300 men, out of which 40 were accepted for the forest battalions. Several of our state forestry men joined these units, which when completely filled and equipped, sailed for the European continent. One battalion went to France, and I heard afterwards that the French people living in the town where it was stationed literally wept tears of anguish when they saw the speed with which their beautiful woodlands, carefully tended for centuries, fell beneath the axes and saws of this forest engineer battalion. However, the needs of the French Army and of the nation for wood and timber were as pressing as those of its allies, so the ancient trees of France were felled and sawed to feed the maws of war.

As a final point of interest concerning World War I, former Forest Commissioner Raymond E. Rendall served in the U.S. Forestry Regiment that had its operations in the forests of France.

In World War II, Maine forests were again considered vulnerable to possible sabotage by subversive agents working within, and by the added menace of air attack and danger from submarines lurking off the coast. While the forests were not considered primary targets, their burning could serve as a diversionary action.

The Maine Forestry Department, unknown to many, played an important role in the national defense effort of World War II. Of particular importance was the strengthening and maintaining of observation posts within the M.F.D. over a territory that could not be adequately handled under the Volunteer Civilian Program.

The Department cooperated with the U.S. Forest Service under the direction of the U.S. Army by implementing a plan of recruiting personnel and converting some existing facilities, towers, and camps, for year-round occupancy on a twenty-four hour daily basis. In addition ten new camps and nine towers were built to augment the existing facilities, thus providing a fairly complete coverage of the forested areas. The surveillance system was supplemented by the Volunteer Civilian Plane Observation Program. Monies for this purpose came from state defense funds and were expended in maintaining four Civil Air Patrol Cub planes, based at Greenville, Portage Lake, and Cupsuptic.

In 1942, the Department's Beechcraft flew one hundred and thirty hours and the four Cub planes six hundred and forty hours, making a total of seven hundred and seventy hours. Two of these Cubs were equipped with skiis for the winter work of inspecting, maintaining, servicing, and administering to the A.W.S. posts. It was no small task to maintain these observation posts on three eight-hour shifts each day, year round. In many instances it was necessary to provide them with supplies over deep snow by tote teams or back-packed by crews on snowshoes.

In the actual operation effort, all observation posts (camps and lookout towers) were given a code name and number. Observers were given special training in plane identification. All aircraft seen or heard were reported directly to a filter center in Bangor. Such calls had a priority for line clearance over all other telephone communications. There were many simulated practice exercises to keep the observers alert and to test their ability to make quick and accurate identifications of aircraft.

Chief wardens and supervisors worked together in the overall administration and inspection of the observation posts, the telephone lines, and the switchboards. After closing the A.W.S. at the end of the war, the Forestry Department reached a residual settlement with the various federal agencies involved concerning facilities and equipment, much of which has long since been phased out.

As World War II continued, a master plan to meet the possible threat of wide-scale sabotage on the forests was worked out with other protective agencies such as the Red Cross, the Office of Civil Defense, State Guard, the F.B.I., First Service Command, Army and Navy authorities, and the Civilian Air Patrol and its auxiliaries.

Still another realistic phase of the national defense effort was the prohibition of any blueberry burning along the coast, which might silhouette the shoreline and give aid to enemy submarines lurking nearby. Eight critical areas, six of which fringed the entire coastline of the state, were mapped and designated to be in need of special fire protection.

In addition to the direct war effort, the Department, in trying to keep up its regular fire protection program, was subjected to curtailment of funds for purchasing equipment and supplies. Federal

AD-126-1  UNITED STATES DEPARTMENT OF AGRICULTURE

Forest Service
INITIATING BUREAU OR OFFICE

Philadelphia, Pa.
LOCATION

Name   Mr. John E. Mitchell   Date   February 22, 1943

NATURE OF ACTION: Change in Status

	From	To
Position	Agent (Aircraft Observer)	Agent (Chief Aircraft Observer)
Grade & Salary		Unallocated, $160 per month
Bureau		Forest Service
Branch		Region 7
Headquarters		To be assigned by Bureau (within State of Maine, Region 7, U.S. Forest Service)
Departmental or Field		Field

Effective Date: March 1, 1943

Remarks:

This action is subject to the provisions of paragraphs   listed on the other side
of this notification.

By delegation of the Secretary of Agriculture in accordance with the authority granted
by the Act of June 26, 1930, Public No. 441, amending Section 169, R. S.

_____
Personnel Officer
TITLE

PERSONNEL NOTIFICATION

ration regulations on tires, batteries, gasoline, and food, posed a special problem in the normal operation of the M.F.D. In addition, personnel left the Department to join the armed services. To offset this drain on vital manpower, key people who had reached the age limit of state employ were granted an extension of time by the governor and council authorization. When the war ended in 1946, the Forestry Department reinstated fourteen men and employed thirty-five veterans to fill in the personnel gaps caused by the war.*

One last item pertaining to World War II deserves special mention. Not generally known is the fact that there were four prisoner of war camps in Maine, one in Hobbstown (T4,R6, B.K.P.-W.K.R), Somerset County; one in Seboomook (T4,N.B.K.P.), Somerset County; one in Houlton, Aroostook County; and one in Princeton, Washington County.

The establishment of these camps came about in a rather interesting way. The war effort called for a great increase in paper products, which meant greater wood production. At this time several representatives from Maine were serving on the War Production Board. They were well aware of the situation and in particular that Hollingsworth and Whitney (now Scott Paper) was the only company in the world making tabulating card stock, an item which was much in demand by the U.S. Army. There being such a prime need, the prison camps were assigned to Maine in an attempt to alleviate the critical manpower shortage in the woods.

The Hobbstown P.O.W. camp was made up of the elite German Africa Corps under the command of General Rommel (the Desert Fox). The total complement of prisoners, officers, and foresters numbered about 300. The prisoners were treated under strict adherence to Geneva Convention rules. Fresh milk, fruits, and sweets were served daily by special delivery service from Waterville, Maine.

These P.O.W.'s came from overseas by boat to Boston, transferred by railroad to Bingham, Maine, and then by truck to the Hobbstown Camp. Many thought this was really the end of the world location. They were allowed to cut only six-tenths of a cord per man per day and were paid 80¢ a day. For the period 1944 to 1946, a total of approximately 34,000 cords of pulpwood were cut.

A letter from Norman Gray of Fryeburg, describes the Seboomook prisoner of war wood-cutting operation:

> The Great Northern Paper Company was granted permission to use prisoners of war in their wood-harvesting operations due to the labor shortages. Upon learning of this, Mr. Roy Wilson of Millinocket was placed in charge of providing the physical unit. This was effected by rebuilding a portion of the Company Seboomook Farm into a Prisoner of War compound, Army personnel housing unit, and civilian housing unit. This was accom-

---

* Honor rolls of those who served in the armed forces are listed in several Department Directories for the period of 1942–46.

plished to Army specification for the P.O.W.s by using one of the larger barns and making the basement into toilet and laundry facilities. Two additional floors were added, making a facility to house a total of about 300 men. An adjacent structure, originally built for a potato house, was modified to an efficient kitchen and dining area. These two structures together with a few acres of land, were enclosed with security wire, full lighting, and guard stations.

The Army Personnel quarters were located adjacent to the Compound and were provided by updating the former Farmhouse and offices. This building housed about 40 military personnel, made up mainly, over the period of the project, of men who had previously served in combat areas.

The Civilian personnel were housed in what was orginally a store located at the Seboomook boat pier—just westerly of the famed Seboomook House.

During the period of making the physical plant ready to receive the P.O.W.s and Army personnel Mr. Lloyd Houghton of Bangor and Mr. Norman H. Gray of Fryeburg were equally active in locating Pulpwood stands for cutting. Mr. Houghton had had many years of association with the pulpwood operations of Great Northern Paper Company, and perhaps more important, was acquainted with many fine older camp bosses or foremen.

Upon arrival of the P.O.W.s at the Seboomook Compound the men necessary to operate the physical unit were located among them. This actually required many men, including cooks, building custodians, electricians, ground keepers, etc., for within both the Compound and the Army areas.

The P.O.W.s available for actual woods work were assigned to work in units of 25 men, including an interpreter. The Great Northern Paper Company had a civilian foreman and straw boss. The P.O.W.s left the compound by 25-man units at 7:00 A.M. each work day and were transported by G.N.P.C. canvas-covered trucks to the work site. Guards rode with the civilian truck driver and an army personnel carrier followed each convoy. In the early stages of the woods operation, the first activity was to clear a one chain strip in a ten chain pattern. After such work was completed by a 4–6 man cutting crew under close security, guards were stationed at each corner and workmen were allowed freedom of foreman-directed cutting operations within the grid. Early security was intense, but after several months, especially during the winter hauling-off season, the workmen were at liberty to move as directed by the foreman within the ground area of the operation.

In the early stages of cutting operations a great amount of time was given to each unit by actual "stump" speech instructions and demonstrations. Very few men had ever used an axe to any extent and the bow saw and cross-cut saw certainly were foreign tools. Considerable time was spent in stump-cutting pulpwood within the grid area to enable detailed instructions.

A pre-allotted number of saws and axes were given out each morning per unit and were checked back each P.M. The hand tools were all handled through the tool repair headquarters set up just outside the compound gate. In the early stages of using axes and saws the breakage of axe handles, axe heads, and saw blades was unbelievably high. It was the feeling of some of the civilians that perhaps some breakage was with purpose. This was overcome by having a sufficiently large crew in the tool shed to always have a full supply of tools ready each morning. Some days this meant hanging up to 100 axes and replacing several dozen saw blades. The workmen soon found they had rather work with new, well-cared for tools than do [an] equal amount with beat-up axe heads and saws with missing teeth.

Generally speaking, in [the] Pittston Farm to Burbank area much of the forest stand and topography was best adaptable from which to cut yarded pulpwood. The Army was not experiencing any problems with security, so we were permitted to use yarding crews. Many of the P.O.W. replacements were Austrians, who were in Rommel's African Campaign. Many of these men were of Farm background and were fine workmen. They took pride in producing an attractive pile of pulpwood. They used the yarding horses with care and did not hesitate to do work themselves that would lessen the burden on the horses.

Each unit for yarding wood consisted of the same 25-man units—6 felling—6 horse teamsters—2 men rolling—2 men sawing, using a 2-man Mall Power saw with 2 men piling. The early Power saws developed many problems when used in P.O.W. production and daily quotas. It became a major task to provide at least one in running condition during the entire work day. This was accomplished to a reasonably satisfactory condition by setting up a saw shop, using 2 or 3 P.O.W. mechanics supervised by a civilian mechanic. Each yarding unit had at least two saws on the yard, plus we employed a mechanic in the woods that had a horse and scoot on which was carried extra saws and standard repair items.

The early quotas were comparatively small, but following training periods, etc., the 25-man unit attained a cord per day per man, either as stump cut or yarded wood.

On this and following pages are German prisoner of war photos taken at Princeton Camp in 1944. These men were from General Rommel's elite German Africa Corps

I was not familiar with the U.S. Army—G.N.P. Co. financial arrangements, but seems to me the G.N.P. Co. may have paid around $5.00 per cord. I think I am correct that the P.O.W.'s account was credited 80 cents per day—10 of which he could use to buy books, etc., for his personal use.

There were many well-educated technicians, specialists and special skilled men in the outfit. The compound electrician was Top Man in Rommel's African Communication system. In an organization of this size, there often developed failures or break downs, but [we] were usually able to make corrections by taking advantage of P.O.W. training. Many of the men continued studies. Some repaired watches and clocks which were made available to them by civilians through the Army.

The general impression was that some of the early men found it hard to adjust out side [a] Military regime. As replacements arrived with men from occupied countries they were willing to do assigned tasks. Some of these men had been P.O.W.s seven years before arriving in Maine. They had very limited knowledge of the fate of their families and were frank in admitting that if to cut pulpwood well might help in their return to their homeland they would wish to remain right in the woods instead of months in various compounds during reparation.

A letter from Robert Leadbetter, of Bangor, addressed to me in March 1976 recalls more details of the German war prison camps at Seboomook.

RATOR BUILDING | CELL BLOCK | INFIRMARY | SENTRY TOWER | ICE HOUSE | RATION BUILDING | STAFF HOUSE | P.X. AND RECREATION BUILDING
SENTRY BOX

OF WAR CAMP
CORP
'46

It was just good to hear your voice again the other day, even though it was kind of second hand via telephone.

I guess at our age, or I'll speak for myself, my age, it just doesn't matter much what we say as long as it comes from the heart. I'm beginning to realize that it takes a lifetime to form definite opinions and to feel that one knows whereof he speaks. In short, I'm satisfied that finally I know whereof I speak and I just want to say, you're a good man McGee!

I'm sure that your life contributions of honesty, integrity, energy, common sense and leadership to the State of Maine and its outdoor people have cut a big and lasting swath! Enjoy your earned retirement!

Just reminiscing a little, I can remember meeting you in 1936. or 1937 when your plane plunked down in 5th St. John Pond, not long after a fire broke out in our pulpwood works. You were the first state man on the job and at the time we wondered how a guy from as far away as Augusta could show up so far back in the woods so soon? If my memory serves me correctly, I believe Bill Turgeon was your pilot.

Funny thing though, I'm really having a tough time recounting many of the particulars which took place about seven years later at the Great Northern Paper Company Seboomook Prisoner of War Camp. However, Tom Russell, a good friend and Company man who was there at the time has supplied a lot, or most of the following information. . . .

Our German prisoners came to the State of Maine in March, or April 1944, after having been employed as cotton pickers somewhere in the South. They were transported by rail to the Moosehead Lake country and disembarked at East Outlet, where they were then trucked to Seboomook, or Northwest Carry.

The prisoners numbered about 200 at the start and later were increased to better than 250. They were housed in two large horse barns which were on location, along with other buildings which made up a Great Northern Paper Company Depot and supply farm for woods operations.

Of course the barns were fitted out and made livable, and high, double barb-wire fencing was constructed around a spacious compound, supported by four guard towers with machine guns.

A two-story potato-house was converted to a kitchen on the lower level with dining room on the upper floor. New barracks were constructed to complement the guards, and U.S. Army officers were quartered in an adjoining Company farm boarding house.

Shortly after the arrival of the prisoners, personnel from the U.S. Forest Service out of Philadelphia were dispatched to instruct the men in the proper use of the axe and cross-cut saw. Axes were sharpened and saws filed by prisoners at their compound.

The Great Northern Paper Company housed their own crew, consisting of a superintendent (L. E. Houghton), five foremen, a cook, cookee, clerk, mechanic and bull-cook in newly constructed quarters in a location between the prisoner compound and Moosehead Lake, some distance from the Army enclosure.

The prisoners were driven to and from a work site at Seven Mile Hill on Boyd Town in 10-wheel trucks (later by bus), a distance of six, or seven miles one way. Later another work site was opened up on Burbank Township, about the same distance from the compound.

Initially about 30 men starting cutting and yarding operations to haul roads which were layed out by Great Northern foremen, and as time went on the prisoner crews were increased so that by the end of the first year they had produced in excess of 12,000 cords, averaging some 275–300 cords per week for the weeks worked.

I believe a Captain or Major Ryder was in charge of the war camp and after operations got under way he notified the men that they would be required to cut one-half cord of wood per day per man.

The men started out using cross-cut saws and production was down, but soon they were instructed in the use of a buck-saw and slowly their production increased. Soon the prisoners were working singly, instead of in pairs, cutting at the stump as our Canadian cordcutters did and they had no trouble meeting their half-cord per man per day quota. Later the daily man quota was increased to three-quarters of a cord.

After the first six or eight-month operating period, Lloyd Houghton was transferred to Houlton where he organized another prisoner woods operation. Prisoners in this new location were housed in Army barracks at the Houlton Airport and were trucked to a pulpwood cutting site in the Haynesville Woods.

I'm not sure how monetary arrangements were adjusted between Paper Companies and the U.S. Government for wood production, equipment, etc., but I seem to recollect that the prisoners were paid 60¢ per day and the Great Northern was assessed at the rate of $4.50 per cord.

It was rumored that prisoners impounded at Seboomook were more or less the cream of the German Army, the nucleus of Rommel's Africa Corps. I will say that outwardly they took their incarceration like men. They were good workers, helped each other produce their daily wood quota, and were respectful at all times to our Company woods personnel. Many of them were farm boys and they understood hard work. They loved horses to the extent that many of them used their noon lunch and rest period to card down their companion working animal. They were very particular that the two horse haul-off sleds were not overloaded and that their team did not develop a collar, or harness chafe.

These boys were not trouble makers and outwardly were not the cruel Nazi fanatic that we often read about during that period. Now and then a bad apple surfaced and if he couldn't be disciplined he was shipped out to another camp where they catered to his kind. This happened a few times at Seboomook.

I remember one time a couple of the more rabid prisoners managed to avoid close woods supervision from their guards long enough to wire a long pole bearing a make-shift Nazi flag to the top of a big spruce tree. When it was later discovered the commanding officer of the camp assembled the whole outfit in an effort to find out who was responsible for such a deed. Nobody confessed and nobody squealed, so that night the entire woods crew was assembled and marched approximately six miles to the flag site where orders were given to cut down the spruce tree, saw it up into pulpwood, burn the Nazi flag on the spot, and then march back to the compound. It was bitter cold and it took most of the night in the open to carry out the commanding officer's order. They all went to work next morning as usual.

During the time that I was there on the job a lot of humorous circumstances surfaced. I remembered watching two rugged boys trying to load a big, ice covered spruce butt on a sled. They just couldn't handle it and walked off in disgust, recommending that it be left for the Russians who they were sure would be the next war prisoners.

Another time the military people called our Company foreman camp, saying they were completely out of butter and might they borrow a couple of cases to tide them over? Fortunately we had a good supply of oleo on hand and were more than willing to help, stating that we'd send it right over, but no, they'd send their own pick-up after it. Half an hour later we received a call asking where the butter was? You just picked it up! Hell, that wasn't butter, that was oleo! Who ate butter during the war years? Evidently they did, but we didn't.

Big steaming kettles of vegetables and meat stew, hot home

made bread and German pastry—the best of good plain food—was trucked out to the operation site every noon where the prisoners gathered around a woods fire and apparently enjoyed themselves. Our Company foremen pulled a peanut butter, or canned corn beef sandwich out from the back of their shirt and were lucky if they had tea to wash it down with. We were not permitted to fraternize with the prisoners on the job and of course the prisoners took orders only from their guards.

Pulpwood cutting prisoners worked six days a week, about seven hours per day, exclusive of lunch hour and travel time. I believe the war camp closed down in the early spring of 1946. The second year pulpwood production was about equal to the first years cut.

Although prisoners had ample opportunity to escape, it would have been a suicide act with only one road leading into an area surrounded by dense woods for many, many miles on all sides. No escape attempts were made.

There are a few knowledgeable old timers still around if you need more help relative to information pertaining to early woods operations in the general North, South and Main River Branches of the Penobscot. Try Spotty Leavitt in Old Town, Felix Fernald and Tom Russell in Greenville, George Hall and Eddie Lumbert in Millinocket, and there are others.

The statement quoted below, written by George Cook, gives some interesting details concerning the prison camp at Princeton, which was a joint effort of the St. Croix, Oxford, Penobscot Development, Hollingsworth and Whitney, St. Regis, and the Atlas Plywood companies, in cooperation with the Federal Government and the army.

REPORT ON THE P.O.W. CAMPS

I received your request for information on the P.O.W. camps. First Georgia-Pacific was not involved in this project in 1944 & 1946. Rather it was the Eastern Pulpwood Co. of Calais, Maine, a subsidiary of Saint Croix Paper Co., Woodland, Maine. Georgia-Pacific bought mill and all holdings in the sixties.

You mentioned in your letter four companies; there were five: Scott Paper, Oxford, St. Regis, Hollingsworth & Whitney, and Eastern Pulpwood Co. Each project was numbered Eastern Pulpwood Co. I was foreman of project No. 5. The P.O.W. camp was situated in Princeton, Maine. The prisoners were picked up by company trucks at the camp at 6:00 A.M., six days a week and taken to different locations, each Co. looking after their own. I do not have the names of the foremen for the other companies as we were quite some distance from each other.

The majority of these companies had the wood cut and piled in small piles in the woods; as was called stump cut wood. Our project, No. 5, was all yarded wood with horses. I worked five men in a crew. We used axes, pulp saws, and crosscut saws with the P.O.W.s I had 40 cutters and one P.O.W. interpreter. If one man was injured a replacement was sent to me the next day. My car was sent to the camp with the injured man as soon as the accident happened. I also had one American soldier as guard, one bus driver, one filer who kept the saws and axes sharpened. The bus driver went out with my car to the P.O.W. camp at 10:00 A.M. and brought hot lunches for prisoners. The interpreter was paid $.80 per day and the prisoner quota was 8/10 of one cord of wood for each man. This amount was required from each before they left the woods. We averaged more than this amount after a few months and they were paid accordingly, maximum was $1.15 per day. This was given to them in a book in stamps and then turned into cash when they went home.

I scaled this wood before the crews left at night and sent a copy of each prisoner's scale to Major Murphy at the P.O.W. camp each night and the monthly scale to Eastern Pulpwood Co. and Major Murphy at the end of each month. The Eastern paid the army $5.00 per cord for cutting the wood and the army paid the prisoners.

This project was started in June of 1944. A sergeant in the Army was foreman until August when I was transferred to this operation. I laid out the truck roads before cutting and each crew had one area marked for cutting; 3–5 chains on each side of the truckroad. This area was set up to last a crew one week. I had one horseler who took care of the horses and brought them to the woods in the morning and took them back at night. I also had 5–6 men cutting roads ahead of prisoners. When one road was finished we bulldozed it and the company trucks took the wood to the Woodland Mill a distance of 30 miles. The mill was short of wood and our holidays were Christmas day and the day after.

This seems to be all I can remember at the present time. Hope you can get something from this.

Figures are lacking on the total wood cut by the P.O.W.s at the Princeton camp, but four thousand, five hundred cords are known to have been scaled for the Oxford Paper Company.

Having dwelt in some detail on the problems as well as the special programs necessitated by the two World Wars, we now turn to another matter that directly concerned the M.F.D. and resulted in the construction of camps of quite a different sort than those discussed and the special employment of large groups of men within the forests of Maine. We are referring, of course, to that period of the great depression and the Civil Conservation Corps.

A complete and detailed account of the C.C.C., in so far as that corps served in Maine, has never been written. References are to be found in the biennial reports of the forest commissioner, daily newspapers, files in the office of the Sewall Company of Old Town, Maine, the records of the U.S. Forest Service, Washington, D.C., and the scrapbooks of the Maine Forestry Department.

The object of the original act that created the C.C.C. was "to relieve distress, unemployment, restore depleted natural resources, and advance an orderly program of public works." In Maine the situation was optimum for just such a program, for President Roosevelt extended the provisions of the act to include services to private land, but only for the purpose of doing work in preventing and controlling forest fires, insect attacks, and tree diseases, and in flood control. Special agreements of understanding were drawn up between the large private corporate landowners, the State Forestry Department, and the U.S. Department of Agriculture and its forestry service.

Thus began for Maine, and especially for the M.F.D., a program of forest fire protection that proved to be most beneficial. The popularity of this program was evidenced by the fact that while the original congressional Act of 1933 was to have expired in 1935, it was extended to June 30, 1937, and then once more for an additional three-year period, and finally ended only in 1942.

By executive order, the President set up an advisory council consisting of members from the Departments of War, Agriculture, Interior, and Labor. The Department of Labor selected men to be enrolled; the Department of War examined and passed on the physical condition of the men, provided clothes and housing, and operated and administered the established camps. The departments of Agriculture, Interior, and War selected and supervised work projects and also furnished tools, supplies, and equipment.

In Maine the initial quota of men recruited at Fort Williams, Portland, Maine, was as follows: 275 experienced men, 1,225 eighteen to twenty-five year-olds, and 150 veterans, for a total of 1,650.

Camps established on private land were directly under the supervision of the forest commissioner in collaboration with the War Department. Twenty-seven camps had been located in Maine, of which eight remained at the beginning of 1941. By 1942 all camps were closed down. Of particular interest to this account are eight two-hundred-man camps and the eleven fifty-to-one-hundred-man side-camps, which were established in the Maine Forestry District.

Mr. James W. Sewall of Old Town, Maine, later of the Sewall

Top: Beddington CCC camps, 1933 — permanent barracks were built later. *Middle:* Flagstaff CCC camps, 1935. *Bottom:* Millinocket CCC camps, 1935

SCHEDULE OF CIVILIAN CONSERVATION CAMPS IN M.F.D.

Location & Number	Personnel	Abandoned	Side Camps[4]
M.F.D.			
Rangeley P-55	Young men	1937	60-man camp, Toothaker Brook, Wilsons Mills
			50-man camp, Otter Brook, Cupsuptic
Flagstaff P-56	Young men	1935	60-man camp, Jim Pond, King-Bartlett Camp
			50-man camp, Bog Brook, Dead River
			24-man camp, Sandy Stream
Greenville P-57	Young men	1938	65-man camp, Shirley, on East Moxie
			65-man camp, South Inlet, Katahdin Iron Works
Seboomook P-58[1]	Veterans	1934	
Millinocket P-61	Young men	1935	96-man camp, Togue Pond, Mt. Katahdin
Patten P-60[2]	Young men	1937	99-man camp, Seboeis on Grand Lake
Beddington P-54[3]	Veterans	1937	50-man at Nicatous, Hancock County
Princeton S-53	Young men	1941	40-man camp, Seavey Ridge, Clifford Lake

[1] Moved to Grant Farm.
[2] Moved to Hay Lake, T.6 R.8, W.E.L.S.
[3] Originally on Airline Road, later moved to Deer Lake, T.34 M.D.
[4] Side Camps were mostly for truck trail construction

Company, was the state coordinator of the program. Jim Sewall was known as one who often cut across the red tape procedures of the government in his zeal to get projects moving. He was apt to act first and discuss later, much to the consternation of Washington officials.

As forest commissioner, I was property custodian for the C.C.C. camps located on private land within the state.

Principal among the projects carried out by the side-camps was that of construction of truck trails. The justification for such road-building was that the back country would thus be opened for quick access by fire fighters and fire fighting equipment. Under federal regulations such roads were built to the allowable maximum width of sixteen feet. When the C.C.C. program was discontinued and these truck roads reverted to the private landowners, a legal question arose as to their continued public use while also being utilized for hauling of logs or pulpwood, for the width did not meet state highway regulations. It was ruled that since public funds had been used in the construction of these roads the public could travel over these roads, but only at their own risk. A system of turnouts was often introduced to take care of trucks and increase safety.

The table on page 191 gives the location along with other pertinent information including the relocation of C.C.C. camps operating within the territory under the protection of the M.F.D.

An emergency side-camp of fifty men from Princeton was located twelve miles from Portage Lake in 1938. The purpose of this camp was to aid the Forestry Department in an operation against the European spruce sawfly. Over one and a half million parasite cocoons were collected by the men in this camp and shipped to the state's insect laboratories at Orono and Bar Harbor, where they were used for breeding purposes.

The following tabulation lists a number of the projects which were accomplished by the C.C.C. program during its operation in the State of Maine:

Bridges, foot and horse	107
Bridges, vehicle	240
Lookout towers	6
Shelters, trail and picnic	81
Telephone lines—miles	482
Truck trails—miles	246
Foot trails—miles	495
Horse trails—miles	137
Firebreaks—miles	13
Fire hazard reduction, miles roadside	551
Fire hazard reduction, in acres	2,198
Man-days forest fire fighting	29–30,000
Man-days in patrolling	1,766

While it can be seen from the table above that the M.F.D. was the beneficiary of numerous physical improvements and construction projects, it was in the area of actual fire suppression that one of the major contributions by the C.C.C. was made. It saved the state well in excess of $70,000 in payroll, representing thousands of man-days in the fighting of forest fires. While the C.C.C. crews were designated as a second line of defense in fire fighting, they more often became the first line of defense in the suppression action. After the C.C.C. program ended, it was surprising to learn of the number of people who responded to the call for forest fire fighters whose experience, learned while members of the C.C.C., qualified them for certain positions in the fire suppression organization. Many produced certification cards as crew boss, pump operator, timekeeper, etc. Thus it became evident that the training provided by M.F.D. fire wardens during the years of the C.C.C. program had been an excellent investment.

Besides manpower, the C.C.C. program provided thousands of hand tools along with back-pumps for fire line construction. In addition, some portable power pumps and one-and-a-half-inch linen hose were also furnished.

A summary of the major contributions of the C.C.C. in the suppression of forest fires within the territory of the M.F.D. follows:

1. Well-trained, organized, and disciplined fire fighting crews. (Who can forget the "one lick method" of establishing hand-dug fire lines developed under this program!)
2. Mass feeding of large crews on the fire site
3. Mass transportation of crews to and from fires
4. Control, orderliness, and packaging of tools and equipment
5. Precautionary measures at all times for safety
6. Marking of hand tools and other fire equipment.

All of these developments and practices served as object lessons in the further development of the M.F.D.'s own fire suppression system.

Three spray planes flying in staggered formation 200 feet above tree level

XI

THE SPRUCE BUDWORM PROGRAM

In all my career since 1929 I have never by far witnessed the grave and extensive holocaust posed to Maine forests by the budworm for 1975.[*]

Just as the Number One natural enemy of the forest, fire, has caused widespread destruction within the State of Maine, so has the Number Two threat—the silent but persistent killer: the spruce budworm.

Although this forest insect has always been present to some degree in forests of the spruce-fir type, an area covering some eight million acres in Maine, it has normally been held in check by natural enemies and the lack of suitable climatic and forest conditions that would favor breeding. However, periodic outbreaks have caused widespread destruction, particularly in the northern area of Maine.

The history of previous outbreaks can be traced back nearly two hundred years. At least six separate and serious epidemics are known to have occurred, in 1770, 1806, 1878, 1910, 1949, and in the late 1960s. Tree growth rings confirm the reports of the earlier outbreaks. Evidence shows a tendency for the movement of this pest from west to east, beginning in Ontario and western Quebec and spreading into northern New York, New England, and the Maritime Provinces.

In his book *History of the Woods of Maine*, Philip T. Coolidge makes reference to the great destruction of spruce and fir by the budworm in a big area east of the Penobscot in 1818 and again in 1880. These infestations caused the loss of one billion feet of spruce and fir

---

[*] Former State Entomologist Robley Nash, from a paper before North American Spruce Budworm Symposium, Washington, D.C., Nov. 1974.

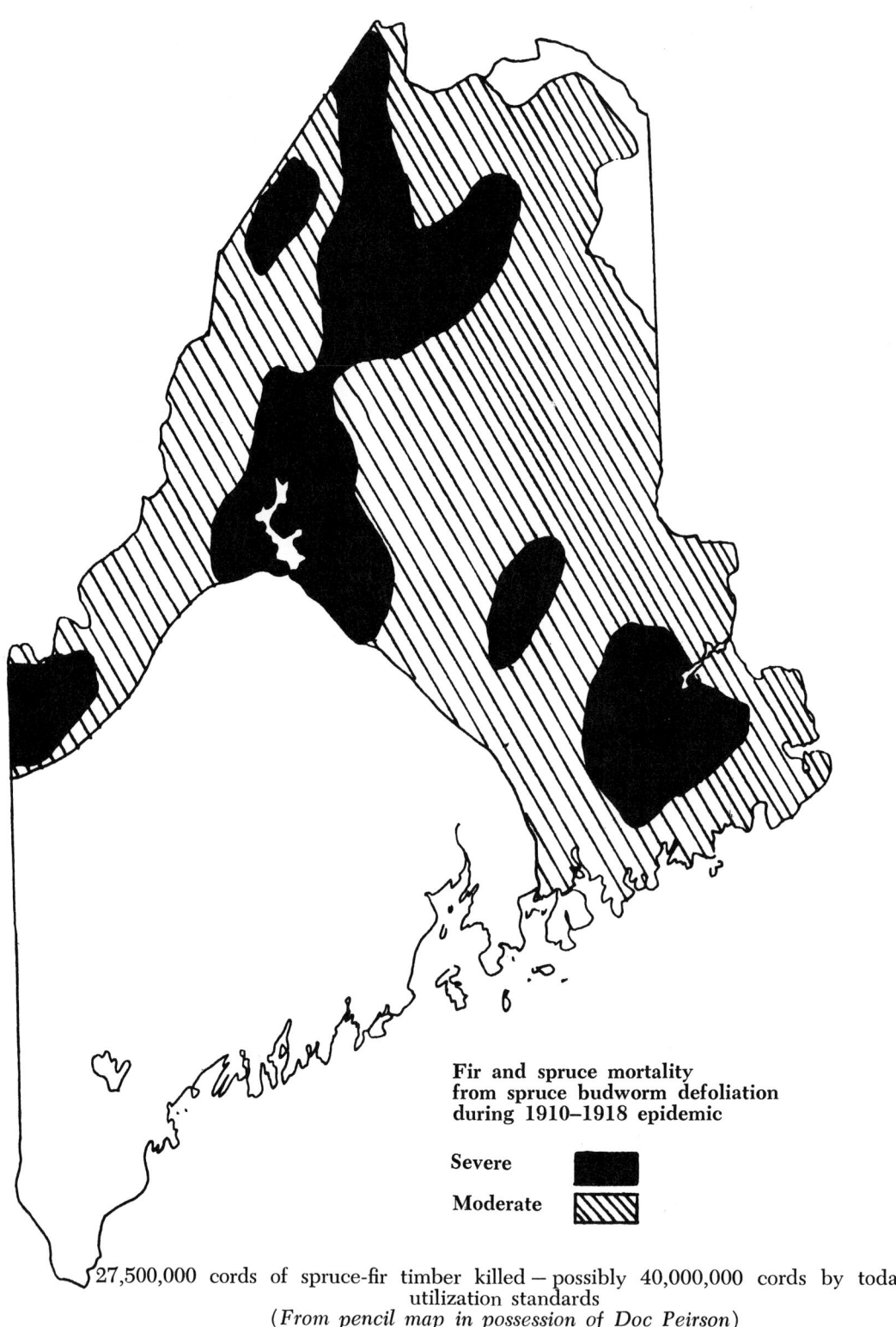

Fir and spruce mortality from spruce budworm defoliation during 1910–1918 epidemic

Severe
Moderate

27,500,000 cords of spruce-fir timber killed — possibly 40,000,000 cords by today's utilization standards
(*From pencil map in possession of Doc Peirson*)

Unsprayed spruce and fir killed by two to four years of severe spruce budworm defoliation

along the Allagash and the tributaries of the St. John, spreading southward into northern Piscataquis County.

The outbreak of 1910-1918 is well documented and is remembered by many observers still alive today. In this epidemic some 27,500,000 cords of spruce and fir were killed in Maine, a figure which might be increased to forty million cords under today's utilization standards. In some townships the kill amounted to ninety per cent of those species of trees attacked by the budworm.

With no action being taken to control the devastation during these earlier epidemics, another threat accompanied the spread of havoc wrought by the budworm. Destructive fires are known to have occurred, fed by the dead and dying trees.

The current budworm outbreak first began in the period from 1949 to 1950, with a tremendous flight of moths descending upon the City of Quebec; and the epidemic has ever since made its eastward advance with an increasing threat of large-scale flight of the adult moth into northern Maine. Presently, Maine is caught in a giant pincer movement with vast areas of spruce and fir infected by the budworm in both the Province of Quebec and New Brunswick.

According to the 1971 Timber Resources of Maine report, seventy per cent of the spruce-fir forest type is found in the four northern counties of the state—Aroostook, Penobscot, Piscataquis, and Somerset. Most of this area lies within the M.F.D. Sizable stands of this type are also found in Washington County and in parts of the remaining counties.

Within this great northern area of the state, the budworm continues to be the Number One insect threat to our forest. Landowners in the District have been and are presently much concerned over the situation and the serious threat to the raw material so vital to the pulp and paper industry as well as to the natural resource of the forests of Maine in general.

The economic impact can be readily understood from the following table prepared for the Environmental Protection Agency in Washington, D.C., as a basis for seeking financial assistance for a major spray operation conducted over an area of 450,000 acres in 1973. It estimated 106,380 cords of spruce and fir already killed in this area and 525,000 cords of standing timber in imminent danger of destruction.

COST/BENEFIT RATIO ANALYSIS 1973[1]

MAINE	Direct Income*	Indirect Income**	Total Economic Impact
Pulp & Paper Mfr.	$8,422,830	$6,670,890	$15,093,720
Lumber Mfr.	2,559,960	2,034,090	4,594,050
Logging & Pulpwood Operations***	1,769,040	1,458,510	3,227,550
Value of Maine Mfr.	12,751,830	10,163,490	22,915,320
Stumpage Value of Logs to Canada	100,000	–	100,000
TOTALS:	$12,851,830	$10,163,490	$23,015,320
Income Per Cord Equivalent	$111/cd.	$87/cd.	$198/cd.

*Direct Income values based upon Maine manufactured product values.
**Using Indirect Internal Multiplier (discounted by 40 per cent for conservative Maine estimate).
***Based on 116,200 cords cut (106,200 cords in Maine use, and 10,000 cords exported to Canada).

[1]Based upon proposed 1973 spray area of 450,000 acres.

Protection against spruce budworm infection is afforded by low-flying teams of spray planes guided and commanded by two small planes flying above them with experienced navigators aboard

Before turning our attention to the large-scale program mounted to meet the present infestation of the budworm, it is necessary to say something about the effort made in confronting this threat in the recent past.

Attention to the problem of destructive forest insects within Maine began in 1921 with the appointment of Henry B. Peirson by Forest Commissioner Samuel T. Dana as the first full-time forest entomologist to be employed by any state. The Entomology Division of the Maine Forestry Department has since grown to include a competent staff of trained entomologists and pathologists with laboratory facilities for both identification and applied research. Robley Nash succeeded Dr. Peirson in the position of state entomologist and was the master-mind behind all the recent spruce budworm spray operations and other related attacks upon destructive pests.

Along with the appointment of a state entomologist, another "first" in the country was the introduction by the Maine Forestry Department of a forest insect survey under the direction of Dr. A. E. Brower. This first took place in the late 1930s when forest fire wardens both in the M.F.D. and in organized towns, as well as forest insect rangers began a seasonable operation of insect collection by "beating trees" under which sheets had been spread and sending whatever was so collected into the Augusta laboratory for identification. Landowners and industry personnel within the M.F.D. also made similar collections.

By means of this field force, it was possible to get good coverage of the forests and to discover new forest insects, to determine the population increase of known insects, and to evaluate the spread and extent of outbreaks. This excellent ground control was integrated with aerial surveys and was especially helpful in understanding the budworm problem.

The Forestry Department also was provided with a statutory provision whereby the forest commissioner can take the necessary control action in cases of emergency resulting from pest outbreaks (Title 12, section 1007, 1964 M.B.S.A.). Such a law is particularly helpful during times when rapid and proper response to dangerous situations is crucial.

It is not the purpose of this narrative to get too involved with all the ramifications of forest pest control. However, the cooperation between the landowners and the M.F.D. on the problem of the budworm must be emphasized. Faced with a joint concern, a proper means of control had to be found. Lack of control not only would result in the direct loss of infected timber, but subsequent danger to the entire forest through the threat of fire.

In the recent outbreaks (1954 and 1974) a number of approaches have been considered and weighed:

1) DO NOTHING—would be (a) to abandon a vast renewable natural resource, (b) leave wide-spread and very intensive budworm populations to threaten larger areas and (c) create an explosive forest fire situation.

2) FOREST MANAGEMENT—different methods of forest cutting practices have had no bearing on the severity of budworm attack during outbreaks of this proportion.

3) BIOLOGICAL CONTROL—was attempted first by the Maine Forestry Department and much effort was expended over the years—without tangible results.

4) CHEMICAL CONTROL—is the only approach that has so far been effective in saving trees from destruction by epidemic budworm infestations. Therefore this was the approach proposed for 1973.

It has been the policy of the Forestry Department to wait the first year or two of a budworm outbreak in the hope that weather and other natural conditions would correct the situation. But when there is treetop branch damage due to defoliation and tree mortality is imminent, then the decision to spray has to be made.

The use of DDT has always resulted in a high percentage of budworm kill. However, due to public pressure by environmental groups during recent years, a ban has been placed on the use of this insecticide. This ban necessitated the search for a substitute. Although not as effective as DDT. the new chemicals Fenitrothion (Sumithion) and Zectran have been used in Maine with continuing pilot studies in other areas of the country.

The following table has been prepared showing all aerial spray operations since 1954, giving the acreage treated, cost per acre, per cent of control, and the insecticide used.

While many may disagree, it has often been stated that the forests belong to everyone and thus the cost of their protection should be shared between private and public interests. Since 1954, it has been customary to carry on the budworm control program on a cost sharing basis. The sharing has been a three-way split: one third paid by the M.F.D., one third from the general state fund, and one third from federal sources.

SUMMARY OF AERIAL SPRAYING FOR BUDWORM CONTROL - M.F.D. and ORGANIZED TOWNS

Year	Acreage Treated		Cost per Acre	Per cent of Control	Insecticide
1954	21,000		$1.54	99	D.D.T.[1]
1958	302,000		.85	96	D.D.T.
1960	217,000		.97	97	D.D.T.
1961	53,000		1.17	98	D.D.T.
1963	750	(test)	-	-	B.T.[2]
1963	479,000		1.06	99	D.D.T.[3]
1964	58,000		1.55	96	D.D.T.[4]
1964	1,108	(test)	-	-	Malathion
1967	500	(test)	-	82	Zectran
1967	92,162		1.60	87	D.D.T.[5]
1968	10,560	(test)	-	-	Fenitrothion[6]
1970	210,000		1.39	84	Fenitrothion[7]
1971	8,736	(test)	-	-	Zectran
1972	500,000		2.71	85	Zectran
1972	200	(test)	-	-	B.T.[8]
1973	450,000		2.71	93	Zectran
1973	20,000	(test)	-	-	Zectran[9]
1974	420,000		2.44	91	Zectran
1975	2,260,400		2.70	98	Zectran
	5,104,416 Acres*			91	Fenitrothion[6]
				91	Sevin

[1] one pound per gallon, per acre, one application
[2] Bacillus thuringiensis
[3] 1/2 lb. per 1/2 gallon, per acre - two applications
[4] Same as above
[5] 87% budworm reduction, 94% overall
[6] Fenitrothion (Sumithion Commercial Trade Name)
[7] Fenitrothion (Accothion Commercial Trade Name)
[8] Bacillus thuringiensis
[9] Different method of applying the insecticide
*This is a cumulative figure for the period 1954-1975 and very little respraying of areas that had been treated previously.

The basis for this cost-sharing lies in the fact that although most of the ownership of land in the M.F.D. is private, the public does have an interest in the great ponds, the fish and wildlife, and in the public reserved lots that are interspersed throughout the unorganized territory of Maine. In addition, an infestation within the forests of Maine endangers the woods within the various municipalities and is a problem to thousands of small private ownerships. Other public interests are the labor markets and industries related to the wood-using mills. Lastly, a federal law provides for financial assistance on just such programs involved in the protection of natural resources.

As a result of such arguments, financial obligations for funding wide-spread insect control have been given to the following agencies by legislative action:

(1) The M.F.D., through funds raised by the mill tax based on a dollar valuation and thus involving all landowners within the District (see below);
(2) Municipalities, through their contribution to the General Fund of the state;
(3) The Federal Government, from funds obtained through application based upon an environmental impact statement as approved by the various federal agencies.

BUDWORM MAINE FORESTRY DISTRICT TAX ASSESSMENTS

Year	Authorization	Mill Rate	Amount
1958	Chap. 424, P.L. 1937	1½	$118,361.31
1960	Chap. 376, P.L. 1959	3/4	65,555.62
1963	Chap. 5, P.L. 1963	2¼	207,386.85
1967	Chap. 101, P.L. 1967	1/2	48,200.16
1970	Chap. 533, P.L. 1969	1	106,104.23
1972	Chap. 617, P.L. 1971	2 3/4	366,612.61
			$912,220.78

NOTE: These assessments are based upon the 1/3 cost share of the M.F.D. for aerial spray operations for budworm control. Where exact amounts are not possible, the difference is met from the reserve from previous budworm operations. Conversely, if the rate is above the 1/3 cost share the balance is kept in a reserve account for future emergency budworm needs. Examples:

1967	1/2 mill rate		$ 48,200.16
	Budworm Balance		5,129.84
			$ 53,330.00
1972	2 3/4 mill rate		$366,612.61
	Budworm Balance		30,054.39
			$396,667.00

Under categories one and two listed above, legislative bills have been drafted and printed and public hearings held before funding was enacted. The M.F.D.'s special budworm tax has been imposed on the landowners six times since 1958, amounting to a total of $900,000. (Funds for 1954 came from private contributions matched with state and federal). Unlike the annual M.F.D. forest fire tax, the budworm assessments have been periodic and made only when the situation required such funding.

As one landowner so aptly stated, "The budworm assessment is a form of forest fire prevention. If trees are not protected, a serious forest fire hazard of dead and dying trees will result."

It is not commonly realized that M.F.D. members have been wholly forest-protection oriented. This is evidenced by their cooperative financing in 1921 of the services of the first state forest entomologist; also of a large biological control program mounted against the European spruce sawfly in the 1930s; an extensive research program on birch die-back and regeneration in the 1940s, and finally the spruce budworm suppression projects started in the 1950s. The funds were derived through taxation of all private landowners within the M.F.D., whether their particular holdings were in immediate danger or not.

Due to the amount of tree mortality from the budworm and the imminent danger of still greater spread, a cooperative project was initiated on a pilot basis between the State Forestry Department, the landowners, and the U.S. Forest Service. This program called for timber salvage operations in affected areas and was to be financed by an allotment of $17,000 for each year of a two-year period. However, the project was dropped when a feasibility study proved that the large-scale organization necessary for the undertaking would be impractical.

The spruce budworm menace is the most serious problem effecting the survival of the spruce and fir forests of Maine. It may well be the most serious threat ever encountered.

In the intervening years since 1970, three major budworm spray operations have been conducted in northern Maine. In 1972, 500,000 acres received treatment, in 1973, 450,000; and in 1974, 420,000 acres. While such operations fulfilled the objective of tree protection, they only reduced the overall population of budworms. The situation was compounded when, in July 1974, massive clouds of moths spread the infestation from Quebec and northern Maine to pretty much blanket the rest of the state. Since the female moth has a tremendous capacity for egg laying, it was not difficult to predict a population build-up of enormous proportions, involving extensive areas of spruce-fir forests.

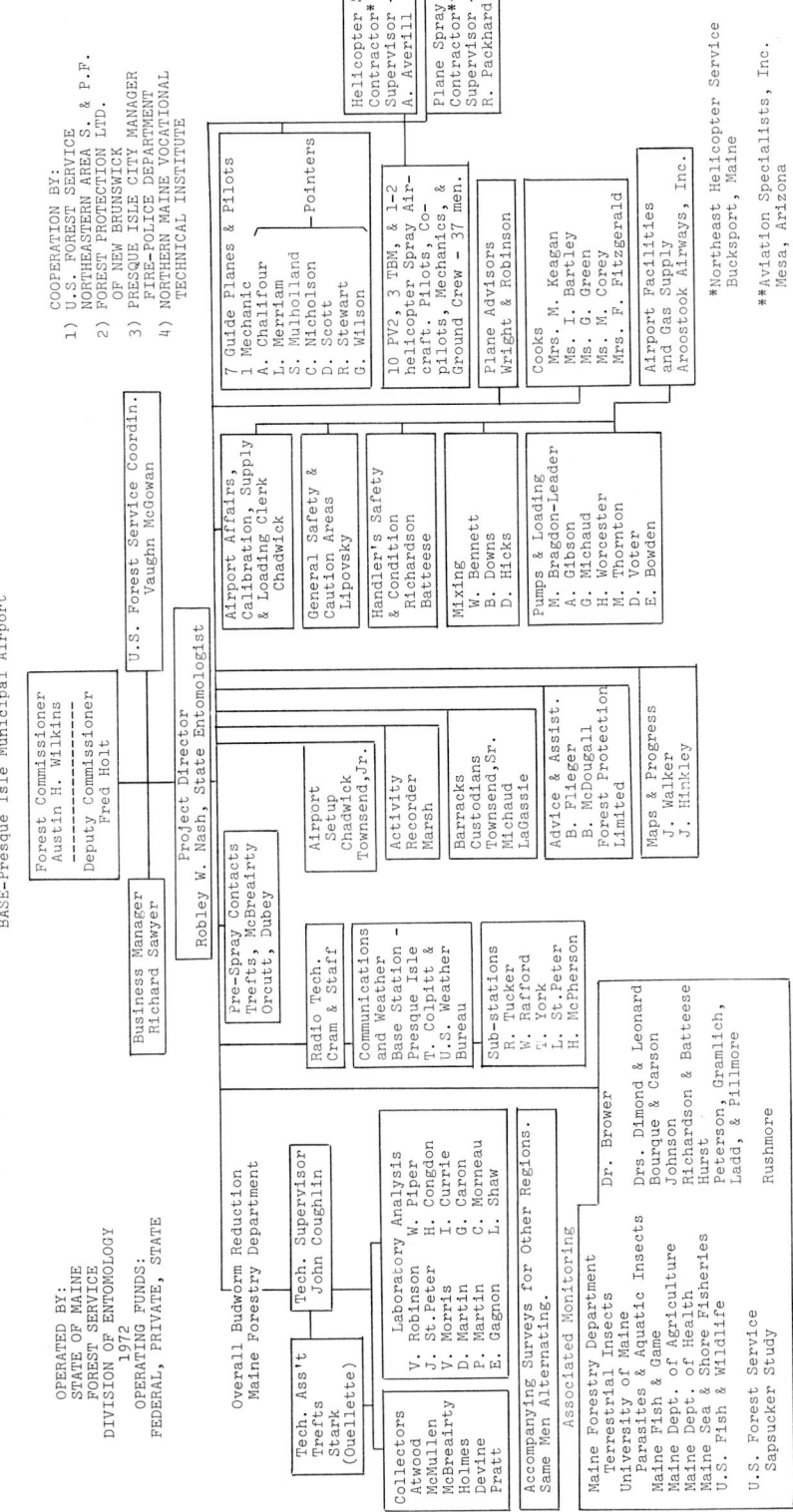

Immediately federal, state, and private agencies began to formulate plans of action.

The problem, of course, did not concern Maine alone. Quebec, in 1974, had already sprayed six and half million acres and New Brunswick, four million acres, in an attempt to curb the epidemic.

(A special word of commendation must go to the late Barney Flieger of Forest Protection Limited, of Fredericton, New Brunswick, who was most helpful in counseling and making available assistance on a number of spray operations in Maine.)

If Maine were to take no control action, the loss estimated for 1974 was 700,000 cords; for 1975, 3,800,000 cords; and by 1980, the staggering figure of 13,000,000 cords. And these figures did not reflect the very substantial decrease in annual growth that would result from the defoliation caused by the spruce budworm.

Under these alarming conditions, the issue changed in 1975 from one of "to spray or not to spray" to one of "who pays to spray," at a cost of $13,125,000 ($3.75 per acre). The traditional procedure of former spray operations was followed, that is, a cost-share basis was set up between federal, state and private landowners.

Money bills (L.D. 620 and L.D. 689) were submitted to the 107th Maine Legislature, and at the same time a deficiency bill was placed before the U.S. Congress, together totaling funds adequate for spraying the recommended acreage.

Several problems had to be faced:

(1) The scarcity of insecticides "Fenitrothion," "Zectran," and "Sevin" had to be dealt with.
(2) The fact that the state's contribution would have to be limited for use only on public lands.
(3) A formula had to be arrived at for private landowner assessment based on the new tree growth tax law in the proposed spray areas.
(4) The question as to the constitutionality of a proposed levy of 30¢ excise tax per acre on all owners of 500 or more acres of land had to be answered.
(5) The raising of an additional research and assessment survey fund of $100,000 was necessary.

The State Supreme Court rendered an opinion that the 30¢ tax was constitutional.

Discussion on the bills ended in a compromise: 3.8 million dollars was appropriated as state and private cost-share to spray 2.2 million acres of the most heavily infected areas, with the state providing one million dollars from the General Fund, of which $480,000 was realized by a built-in "budworm assessment" in the tree growth tax law;

Revised Spruce Budworm Proposed Spray Areas 1975
Estimated 2.2 million acres
Based on availability of insecticides

$2,850,000 from an excise tax upon private landowners; and $3,750,000 as the federal cost-share (restricted to spraying only). The funds for research were granted in addition.

The necessary bills were passed by the State Legislature (Chapter 162, P.L. 1975) and by Congress as an emergency measure to meet the deadline date set for the spraying operation.

As part of an impact statement in a report to the Federal Environmental Protection Agency written to gain approval for the spraying, one finds the following ratios, which are of special interest: Benefit-cost ratios show 2/1 as based on stumpage value, 2.17/1 on stumpage, recreation and fire prevention cost saving, and 32.1/1 using multipliers of mill product and value.

In addition to financial problems, serious objections were raised as to the environmental feasibility of the spraying program. Preliminary injunctions to halt the spraying operation were sought in three suits filed in the Kennebec County Superior Court. After careful deliberation, Justice Edward Stern denied the motions on the basis that one half of the operation had already been completed and that federal cost-sharing was sufficiently assured to finish it. Also that estimated damage to the forest and wood resources should the spraying program not be completed in time for maximum effect, and the potential of increased danger of forest fire from large areas of dead trees, were justification enough.

By June 1975, the spraying of 2.2 million acres was completed. In addition to the application of the insecticides Fenitrothion (Sumithion) and Zectran, an experimental spraying of 12,500 acres with Bacillus thuringiensis (BT), a bacteria that attacks the larvae of moths and butterflies, was tried.

Certainly a major consideration in the decision to spray large areas of forest, and one which directly concerned the program of the M.F.D. was the threat of fire. Given dry weather conditions and acres of defoliated and dead trees, Maine could experience a forest fire holocaust the like of which has never been witnessed. Even with future spraying programs or some other efficient means of budworm control, it may well be necessary for extreme forest fire prevention measures to be observed by everyone to avoid the loss of great expanses of Maine's most valued forests.

New advances are being made in the District in aerial photography, involving various devices, such as that of using infra red film, for detecting and appraising the extent of budworm infestation.

One can only compare the magnitude of preparation for an aerial spray operation on such a large scale with getting ready for a major battle. Some appreciation of the program and logistics involved can be gained from the organizational chart below prepared for 1972.

With a continuing budworm problem it is evident that the emphasis must shift from one of entomological concern only to one including the total effort of forest management. The present situation also illustrates the fundamental importance of cooperation between the M.F.D. and the landowners. The solution must be found in working together, just as it was in creating a program of adequate forest fire protection.

First Annual Meeting of Landowners and Chief Wardens, Augusta, April 8, 1919

1. H.J. Craig, Bingham, representing Kennebec Valley Protective Assn.
2. S.S. Lockyer, Berlin, N.H., representing Brown Company
3. Herbert A. Folsom, Augusta, representing Forestry Dept.
4. J.J. Kneeland, Topsfield, Chief Warden, Dist. No. 3 of Eastern Maine (later merged with St. Croix Dist.)
5. A.P. Daniels, Portland, representing Western Electric Co.
6. Grover C. Bradford, St. Francis, Chief Warden, Dist. No. 3 of St. John Waters (later Allagash Dist.)
7. Fred S. Bunker, Franklin, Chief Warden, Dist. No. 2 of Eastern Maine (later Union River Dist.)
8. Thomas Griffin, Millinocket, Chief Warden, Dist. No. 3 of Penobscot Waters (later Katahdin Dist.)
9. H.B. Buck, Bangor, Honorary Chief Warden, representing Pingree Land
10. Ralph L. Brick, Chesuncook, Chief Warden, Dist. No. 5 of Penobscot Waters (later Chesuncook Dist.)
11. Claude M. Austin, Stockholm, Chief Warden, Dist. No. 1 of St. John Waters (later Madawaska Dist.)
12. Frank W. Hilton, Bingham, Chief Warden, Dist. No. 4 of Kennebec Waters (later Carlin Pond Dist. in part)
13. William Jolly, Bingham, Inspector Maine Forestry Dist. Tel.
14. H.E. Shepard, Bangor, Forester of Eastern Manufacturing Company
15. S.C. Cummings, Haynesville, Chief Warden, Dist. No. 8 of St. John Waters (later Mattawamkeag Dist.)
16. R.E. Pineo, representing American Thread Co. of Milo
17. John E. Mitchell, Patten, Chief Warden, Dist. No. 4 of Penobscot Waters (later East Branch Dist.)
18. John J. Comber, Caribou, Chief Warden, Dist. No. 3 of Kennebec Waters (later Carlin Pond Dist. in part)
19. Henry Cowell, Skowhegan, representing Coburn Heirs, Inc.
20. Neily Violette, Augusta, Deputy Forest Commissioner
21. Blaine S. Viles, Augusta, Honorary Chief Warden (Timberland Owner)
22. Fred A. Lancaster, Old Town, Chief Warden, Dist. No. 6 of Penobscot Waters (later Chamberlain Dist.)
23. Forrest H. Colby, Bingham, Forest Commissioner
24. D.H. Lambert, Old Town, Chief Warden, Dist. No. 1 of Penobscot Waters (later Seboomook Dist.)
25. Charles L. Weeks, Ashland, Chief Warden, Dist. No. 6 of St. John Waters (later Aroostook Waters Dist.)
26. E.C. Hirst, Forester, State of New Hampshire
27. Charles C. Murphy, Rangeley, Chief Warden, Dist. No. 2 of Androscoggin Waters (later Rangeley Dist. in part)
28. James M. Pierce, Houlton, Chief Warden, Dist. No. 7 of St. John Waters (later Dist. No. 9 in part)
29. Chester W. Alden, Westbrook, representing S.D. Warren Company
30. William H. Hinckley, Ashland, Chief Warden, Dist. No. 5 of St. John Waters (later Upper St. John Dist.)
31. Archie G. Norcross, Augusta, Dept. Engineer
32. E.I. Small, Bingham, Lookout Inspector
33. Ervin L. McKenney, Bangor, Chief Warden, Dist. No. 4 of St. John Waters (later Seven Island Dist.)
34. Clyde C. Fox, Wilsons Mills, Watchman on Aziscoos Mountain
35. Harry Davis, Monson, General Deputy, Penobscot Waters
36. N.A. Collins, Boston, representing Western Electric Company
37. A.R. Henderson, Kingfield, Chief Warden, Dist. No. 6 of Kennebec Waters (later Carrabassett Dist.)
38. George G. Nichols, Jackman, Chief Warden, Dist. No. 5 of Kennebec Waters (later Moose River Dist.)
39. A.H. Chase, Milo, Chief Warden, Dist. No. 2 of Penobscot Waters (later Pleasant River Dist.)
40. LeRoy Brown, Lee, Chief Warden, Dist. No. 1 of Eastern Maine (later Passadumkeag Dist.)
41. E.M. Chase, Brownville, former Chief Warden, replaced by A.H. Chase, (No. 39 on list)
42. Frank P. Conley, Greenville Jct., Chief Warden, Dist. No. 1 of Kennebec Waters (later Moosehead Dist.)
43. Frank E. Patten, Cherryfield, Chief Warden, Dist. No. 7 of Eastern Maine (later Union River Dist. in part)
44. A.P. Belmore, Princeton, Chief Warden, Dist. No. 4 of Eastern Maine (later St. Croix Dist.)

208

XII

PERSONNEL AND PUBLIC RELATIONS

Learn to see behind the bark of a tree. Learn to see behind the rough exterior of people. Beneath there is something fine, beautiful and useful.*

The role of the forest commissioner as head of the state's Forestry Department and the M.F.D., over the years, has been of long-reaching importance. The dedication and the services rendered by men who have held this post become evident in their detailed reports. These annual reports, beyond their important documentation, statistics, and historical record, bespeak a deep concern for the preservation of Maine's great forest.

Below is a copy of the first tabulation for the period of 1891 to 1972, showing the chronological record of all appointed forest commissioners and the governors each served under. Research at the State Library also provided the consecutive order of annual and biennial reports submitted by these commissioners.**

REPORTS OF FOREST COMMISSIONERS, TENURE OF OFFICE AND GOVERNORS SERVED UNDER

Year	Forest Commissioner	Governor	Report	Maine Statute Reference
1891-92	Cyrus A. Packard[1]	Edwin C. Burleigh	1st Annual	Chap. 100, Sec. 2 1891 P.L. Maine[2]
1893-94	Charles E. Oak	Henry B. Cleaves	2nd "	" "
1895-96	" "	" "	3rd "	" "[3]
1897-98	" "	Llewellyn Powers	No Report	--
1899-1900	" "	" "	No Report	--
1901-02	Edgar E. Ring	John Fremont Hill	4th Annual	Chap. 100, Sec. 2 1891 P.L. Maine[4]
1903-04	" "	" "	5th "	Chap. 7, Sec. 51 1903 R.S. Maine
1905-06	" "	William T. Cobb	6th "	" "
1907-08	" "	" "	7th "	" "
1909-10	" "	Bert M. Fernald	8th "	" "

* The Reverend J. F. Titus Oates, Episcopal Church, Camden, Maine.
** See Appendix VII for list of the State land agents (1842-1880) who preceded the commissioners.

209

Years	Commissioner	Governor	Legislature		Statute
1911-12	Frank E. Mace	Frederick W. Plaisted	9th	"	"
1913-14	Blaine S. Viles	William T. Haynes	10th	"	"
1915-16	Frank E. Mace	Oakley C. Curtis	11th Biennial	Chap. 8, Sec. 28[5]
					1916 R.S. Maine
1917-18	Forrest H. Colby	Carl E. Milliken	12th Biennial	"	"
1919-20	" "	" "	13th	"	"
1921-22	Samuel T. Dana[6]	Percival P. Baxter	14th	"	"
1923-24	Neil L. Violette[7]	" "	15th	"	"
1925-26	" "	Ralph O. Brewster	16th	"	"
1927-28	" "	" "	17th	"	"
1929-30	" "	Wm. Tudor Gardiner	18th	"	Chap. 11, Sec. 9[8]
					1930 R.S. Maine
1931-32	" "	" "	19th	"	"
1933-34	" "	Louis J. Brann	20th	"	"
1935-36	Waldo N. Seavey[9]	" "	21st	"	"
1937-38	" "	Lewis O. Barrows	22nd	"	"
1939-40	Raymond E. Rendall	" "	23rd	"	"
1941-42	" "	Sumner Sewall	24th	"	"
1943-44	" "	" "	25th	"	Chap. 32, Sec. 14[10]
					1944 R.S. Maine
1945-46	" "	Horace A. Hildreth	26th	"	"
1947-48	A.D. Nutting[11]	" "	27th	"	"
1949-50	" "	Frederick G. Payne	28th	"	"
1951-52	" "	" "	29th	"	"
1953-54	" "	Burton M. Cross	30th	"	Chap. 36, Sec. 17[12]
					R.S. 1954, Vol. 2
1955-56	" "	Edmund S. Muskie	31st	"	"
1957-58	Austin H. Wilkins[13]	" "	32nd	"	"
1959-60	" "	Clinton A. Clauson[14]	33rd	"	"
		John H. Reed			
1961-62	" "	John H. Reed	34th	"	"
1963-64	" "	" "	35th	"	Title 12, Sec.509[15]
					MRSA 1964, Vol. 6
1965-66	" "	" "	36th	"	"
1967-68	" "	Kenneth M. Curtis	37th	"	"
1969-70	" "	" "	38th	"	"
1971-72	" "[16]	" "	39th	"	"
1973	Fred E. Holt*	" "	39th	"	Chap. 460, P.L. 1993

[1] Packard served only one year. Oak appointed commissioner 1892.
[2] Chap. 100, Sec. 2, 1891 P.L. Maine "...report to be made by him annually to the governor on or before first day of December."
[3] No reports for 1897, 1898, 1899, 1900 or 1901
[4] Chap. 7, Sec. 51, Revised Statutes 1903, Maine (Printed in accordance with first Revised Statutes of 1903).
[5] Chap. 8, Sec. 28, Revised Statutes 1916, Maine "...a report to be made by him biennially to the governor on or before first day of December."
[6] Dana served 1 yr. 7 mo. Neil Violette appointed acting commissioner by Governor Baxter.
[7] Violette appointed commissioner by Governor Baxter in 1924.
[8] Chap. 11, Sec. 9, Revised Statutes, Maine 1930 "...report on first day of July to the governor biennially for the 2 preceding years." (Means calendar years)
[9] Violette died September 16, 1935. Seavey appointed commissioner by Governor Brann 1935
[10] Chap. 32, Sec. 14, Revised Statutes, 1944 Maine "...report first day of July to the governor biennially for the 2 preceding years." (Means calendar years)
[11] Nutting appointed commissioner by Governor Hildreth in 1948.
[12] Chap. 36, Sec. 17, Revised Statutes 1954, Maine, Vol. 2
[13] Wilkins appointed commissioner by Governor Muskie in 1958.
[14] Governor Clauson died in office, December 30, 1959. John H. Reed sworn in as new governor, December 30, 1959.
[15] Title 12, Sec. 509, Maine Revised Statutes Annotated, 1964
[16] Wilkins resigned January 1, 1973. 39th report prepared by successor Fred E. Holt, appointed commissioner by Governor Curtis in 1973.
*Title of commissioner ended with creation of Department of Conservation. Holt served as the last Forest Commissioner 2/7/73-10/1/73 (7 months 21 days)
Note coincidence that 5 forest commissioners - Oak, Ring, Mace, Rendall and Holt carried middle initial "E."

Any consideration of the position of forest commissioner must include the matter of salary. It is set by act of the Legislature, which meets every two years unless called into special session. There have been three distinct periods of payment changes. Between 1891 and 1908, the commissioner was paid from General Fund appropriations. Under Chapter 100, section 1. P.L. 1891, the first salary was two hundred dollars per year plus travel expenses, in addition to the monies earned as the state's land agent. In 1909, this was increased to four hundred dollars per year. Then followed a series of legislative bills pertaining to salaries of department heads that involved adjustments within the brackets of "unclassified state officials" with fixed ceilings. These often coincided with state employee raises.

It is interesting to point out that these periodic salary adjustments affecting all M.F.D. employees were the result of special studies to correct certain inequities by establishing a salary plan to remain competitive with other states, to attract the best possible talent, to retain those who wished to continue to stay and work in Maine, and to make salaries more commensurate with their responsibilities. One such study was made in 1968 by Cresap, McCormick, and Paget with legislative appropriation.

A partial schedule is shown below of some of the periodic salary increments granted to the forest commissioner:

1950 —	$ 8,000	1966 —	$14,300*
1954 —	9,000	1967 —	16,500
1956 —	10,000	1968 —	18,000
1957 —	11,250	1969 —	19,500
1963 —	12,250	1971 —	20,500
1965 —	13,000		

* Longevity benefits started this year and have been included with base pay ever since.

For many years the salary of the forest commissioner was divided by statute into payments of one third from the General Fund appropriation and two thirds from M.F.D.'s tax funds. This arrangement was based on the premise that much of the commissioner's time was spent in administering District affairs and that the District's tax should provide an appropriate part of his salary. Only one weekly pay check was issued from the two sources of funds. The process was handled as an internal bookkeeping matter from the comptroller's office.

Then in 1967, the Legislature (under Chapter 476, section 15, R.S., title 12, section 501, amended) decided that the full salary of the forest commissioner should be paid from the General Fund. In all candor, this move was made to eliminate the unfortunate and unfounded idea held within certain circles that with a proportionate share of his salary coming from the M.F.D., the commissioner was unduly obligated to the large timberland owners. This change in salary arrangement resulted in a saving of about ten thousand dollars annually to the M.F.D. and added a drain to the General Fund.

Along with the dedication exhibited by the commissioners, the service of the fire wardens laboring in the field deserves special recognition.

Even with the subdivision of the vast woodlands under protection of the M.F.D., the area within each division remained huge, and in the matter of patrol and supervision called for prodigious feats on foot and in canoes. An early report speaks of one warden averaging two hundred miles per month on his rounds among his patrolmen and lookout tower watchmen. The following quote from a report written in 1923 by Chief Warden John Mitchell gives further evidence of the task involved:

> My territory covers twenty-four towns, or 864 square miles. I visit every patrolman and lookout man at least once a month. To make all the lookout men a visit and return to Patten necessitates a walk of 212 miles. This is by no means all the walking I do in the forest. It is possible that I walk fifteen miles a day on an average during the five months the lookout men are on the job. I have been all over the northern part of Maine through the forest and have not yet been lost.

In 1918, Chief Warden Thomas Griffin of Millinocket reported that he had bought a bicycle to use while patrolling two of his townships—a mode of transportation which was certainly novel in an era that called for woodsmen used to the long trail, the pole, and paddle.

The number of dedicated wardens has become legion during the years of the M.F.D.'s program within the state. It is possible here to mention but a few.* Several members of the well-known Bartlett family worked off and on for the M.F.D. Joshua B. Bartlett, chief warden in 1903 at Ashland; Maurice Bartlett, also a chief warden, and the latter's brother Hugh, who served from 1930 to 1952, were all from this lumbering family. The Weeks family represents three generations of employment: in the M.F.D. Charles Weeks, as chief warden, Ashland, 1917–1939, and his son Harold A. Weeks, who after serving as his assistant, 1935–1939, became chief warden at Ashland, himself, 1940–1958. Dwinal Weeks, his son, was fire warden at Houlton (Organized Town), 1949–1951.

A long line of men served in dual capacity as seasonal chief wardens and off season as employees of Pingree Timberlands (now part of the Seven Islands Land Company). These include Grover Bradford, Harold Pelletier, Stanley Drake, Albert Baker, and John Sinclair, in northern Maine, and Ken Hinkley and O. Lee Abbott, in the western part of the state. John Sinclair, who is currently (1978) president of the Seven Islands Land Company, was a chief warden for the M.F.D., and it was in his district that radio and danger stations were established for the first time. And so the list might go on.

---

* See Appendix VII for a list of chief wardens representing various periods in the M.F.D.'s history.

Following is a list of state professional foresters taken from the personnel files of the Forestry Department. All but two of them are currently in private industry within the territory of the M.F.D.:

Burgess, Sumner	Oxford Paper Company
Clement, David	S. D. Warren Company
Jackson, Wayne	S. D. Warren Company
Ladd, Abbott	Oxford Paper Company
Macomber, Elwin	(formerly) St. Croix Paper Company
Orach, Stephen	S. D. Warren Company
Swenson, Clifford	Seven Islands Land Company
Warren, David	St. Regis Paper Company
Wing, Morris	International Paper Company
Woodsum, Kenneth	(formerly) St. Regis & Huber Corporation

It would not be amiss to mention the warden's wife at this point. She is an unsung figure behind the scenes except for the following tribute paid by one "Skippling, a warden": *

DEDICATED TO THE CHIEF WARDEN'S WIFE

I know a grand person who has what it takes
She stews and she boils, she fries and she bakes,
She sews, cans and pickles, and washes the clothes,
Hoes, mows and irons and stirs up the doughs.
She keeps the Forestry boys happy, with laughter and wit
Also with the public she makes a big hit,
Dishing out permits and weather combined,
Sketching roads, lakes and streams and instructing the blind.
Mapping highways and byways, camp sites and the slash
Tells the sports and the bums where trout jump and splash.
Warns all of the road, it's so crooked and narrow,
Watch out for the log trucks, but follow the arrow:
It points to the pond where this permit is for.
Keep your campfire small and well out on the shore.
Above all, before leaving drown it with care
Stir and mix, mix and stir, 'til no heat is left there.
This protects our wildlife, our trees and our land,
And all nature's beauty from rock slides and sand.
Just one careless camper can cause a disaster,
FIRE is a good servant, but a most ruthless master.
She hello's and ten eight's, goodbyes and ten seven

---

* "Skippling" is a *nom de plume* for Harold A. Weeks. His wife, Crystal, served at his Squa Pan Tower, then as telephone operator at Mouth of the Oxbow Road, and finally as telephone operator at District Headquarters, Ashland, where she also issued fire permits.

And really deserves the best seat in heaven
For spreading, prevention from April 'til fall
And helping keep District personnel on the ball.*

In 1909, with the creation of the M.F.D., the power of appointing chief wardens, wardens, patrolmen, watchmen, and general deputies was given to the forest commissioner. Under Title 12, sub-chapter II, section 521, the law currently in effect still leaves the power to appoint vested with the commissioner, but the language of the law has been changed from "forest fire warden" to "forest ranger." Under the same sub-chapter, section 523, the forest commissioner may also appoint general deputy wardens. In addition the statute, under section 521, has been broadened to include the appointment, subject to the Personnel Law, of other personnel not primarily engaged in forest fire control.**

Original certificates of appointment by the forest commissioner cannot be found. However, it has been possible to trace from some of the earlier certificates issued, the interesting evolutionary changes in the printed forms up to the present time. A close look at copies of certificates issued in 1915, 1918, 1925 and 1935 to Chief Warden John Mitchell are quite revealing. These are on printed forms size 8½″ x 11″. Some contain the printed name of the forest commissioner while others had the name typed in with space indicating the township territory for each appointee. In 1915–1918 the rate of pay was printed *"at a compensation of $3.00 per day, and actual expenses"* and in 1925–1935 it was $4 per day. These certificates of appointment were sworn to before a local justice of peace. In one instance the wife of Chief Warden John Mitchell (Eunice) swore in her own husband. Later on, the division supervisors had the power of attorney and swore in their own warden personnel.

The 8½″ x 11″ certificates were cumbersome to carry around so a change was made to a smaller form which could be carried in a pocketbook and readily shown for identification; sizes were 2″ x 3½″ and later 3″ x 4½″. Finally a certificate of appointment form was printed in pads of 25 with duplicates on 4″ x 7½″. These were designed to be posted at the headquarters of each warden and a special identification card was to be carried in the pocketbook.**

In addition to the regular active wardens, certificates of appointment were made out to honorary chief wardens and in blanket form to railroad patrolmen and general deputies and especially to industry representatives. The latter were people from Diamond International,

---
* Printed in *Forest Protectors*, Vol. VII, No. 3, 1959.
** See Appendix VII for statutory powers stated in the law governing appointments; and for copies of certificates of appointment.

Georgia-Pacific, Great Northern, Oxford Paper, Seven Islands Land Company, International Paper, Prentiss & Carlisle. Names of those to be appointed were sent in to the Augusta office. Later, in 1967, a letter of understanding was used covering railroad section foremen and replacing the annual certificate of appointment.

From the original number of 29 fire wardens appointed in 1891 and 141 in 1903, appointments continued to increase as the fire organization grew, to a figure running into the hundreds by the decade of the 1960s.

In 1909, there were one hundred and forty-one wardens appointed within the M.F.D. Over the years and due to many factors, the overall personnel grew to four hundred and forty-seven in 1925 and to five hundred and twenty in 1952. From this latter date, the employment began to taper off, down to four hundred and twenty-eight, including deputies on call, in 1971. The reduction was the result of cutting back on deputy wardens on call and of the transition from lookout watchmen to contractual or aerial surveillance or detection.

The number of persons employed was not the only factor that changed in regards to personnel with the passing of the years. A significant change affecting M.F.D. personnel occurred under Chapter 147, P.L. 1969 (Maine). This act placed all unclassified employees under the state classified system, subject to rules and regulations of the State Personnel Board.*

Under the former system of unclassified services, the forest commissioner had rather broad powers to hire and fire, create, reduce, or change positions. He also established wages and salaries for all personnel within the M.F.D. organization.* In other areas he had the freedom to deal with grievance cases, minimum delay in appointments, and was under no subjection to rules and regulations of the State Personnel Board, particularly in examinations and eligibilty lists.

It is to the distinct credit of the M.F.D.'s advisory committee that a policy was adopted instructing the forest commissioner to make payrolls of District personnel comparable to the Classified Service of State Employees, including fringe benefits of state employ. This was a wise move. It avoided a dual setup of differential wage scales for similar positions in Organized Towns and in the M.F.D. There were financial problems, for funds had to be found to meet the pay increases decreed by Legislature. During one biennium (1966–67) a twenty per cent wage increase was paid in increments of ten, five, and five per cent. For the M.F.D. payroll this meant an increase totaling over $100,000 for which new funds had to be found. The obvious answer

---

* See Appendix VII for copy of this public law; also for typical M.F.D. personnel salary memo.

was the raising of the M.F.D.'s fire tax rate. This was subsequently done—not just once but several times.

In addition to wage and salary increases was the item of the District's contribution toward retirement longevity benefits for those having eight years of service and for those having over fifteen years, based upon their ages. More recently, was the added cost resulting from the twenty-five year retirement act for all fire wardens who could qualify. An indication of how this retirement item has increased is shown by the following figures for the fiscal year 1972: In that year payment by the M.F.D. was 15.8 per cent with 7.5 per cent being paid by the warden, while outside the warden service, the M.F.D.'s contribution was 8 per cent, with the employee contributing 5 per cent.

Up to 1967, the M.F.D. had never carried insurance policy protection for its regular fire wardens, who might be seriously injured or even killed in the regular performance of their duties. Cases were handled as a budget item when they occurred.

Suddenly it was realized that the District could be in serious financial difficulties should a series of severe or fatal accidents occur within its ranks over a short period of time. Fortunately the Attorney General's office provided the answer by making a verbal ruling that the creation of the Maine Forestry District in 1909 constituted an "established and incorporated administrative district for forest fire protection purposes." On the basis of this ruling, a workman's compensation policy was written for the District in 1967 by John C. Paige Company of Portland, Maine, as agent for the Travelers Insurance Company of Hartford, Connecticut.

A three-year, "retro-respective rating plan" policy was written, providing coverage of $200,000, with premiums of $15,000 to $20,000 payable and adjustable annually within the field limits of the plan. One basic feature was that the insured actually developed his own rate by the relationship of premiums paid to losses incurred. If the loss activity was low, credits would be applied annually and in the final audit. If, on the other hand, losses were severe, the premium was adjusted within the limits of the filed plan on an annual and audit basis. Loss payments to date have not been severe and credits have been made.

This form of policy protection for fire wardens has proved a most beneficial part of the financial program of the District.

While the changes mentioned above were taking place, still another transition was occurring that was of fundamental importance to the employment structure of the M.F.D. Toward the end of the 1960s, it was apparent that serious considerations had to be given to the matter of year-round employment for chief wardens and some other specialized personnel within the warden service. In earlier years, wardens had little difficulty in finding work during the "off fire season." Such jobs were quite varied and included cutting pulpwood, marking timber, scaling, snow plowing and sanding, truck driving, potato inspection, clerking, etc. Whenever possible other agencies such as the landowners and the paper industry were encouraged to employ fire wardens during the off season.

However, as such opportunities waned, chief wardens became increasingly interested in the security of year-round employment. Due to the rising cost of living and other factors, along with the need to hold good men, the M.F.D. started a projected plan to absorb some of the chief wardens on a more permanent basis. The alternative would be a frequent turnover of personnel.

At the time there were twenty-four seasonal chief wardens. The M.F.D.'s advisory committee set a goal of reducing this number to twelve or fourteen and providing these with year-long employment. The procedure followed was not that of replacing chief wardens who were soon to retire, but of combining their districts with others. In this way, two things were accomplished, good men were retained and where vacancies did occur the factor of year-round employment was an attraction for a higher type of warden. In filling vacancies in the new all-year positions, priority was given to promotion from the lower ranks for all those persons who qualified.

The M.F.D.'s advisory committee approved 1969 budget increases permitting twelve chief wardens to work an additional twenty weeks. Initially this was in the form of replacements for those full-time employees who were on winter vacation, and the duties largely involved the repair and maintenance of equipment. The final result was hiring twelve chief wardens of high caliber on a permanent basis.

During the years of rapid change in employment standards and organization, the M.F.D. and the Forestry Department, as a whole, have been relatively free from problems with unions or other groups, as well as from the pressure for political patronage. This is not to say, however, that there was complete immunity.

Many fire wardens are paying members of the Maine State Employees Association. This association has been helpful in promoting and supporting wage increases and other matters relating to the general welfare of state employees. While there have been confrontations with the forest commissioner on grievance cases, all have been conducted in a friendly spirit of discussion.

In addition to the MSEA, a few wardens were members of the AFL-CIO. At one time forty-five to fifty employees of the M.F.D. were dissatisfied with working conditions and sought help from this union. Membership in both the MSEA and the AFL-CIO was made possible through the unique Council Order number 916 (April 7,

1966), whereby the State Comptroller was authorized to make payroll deductions for dues.

Secretary of State

State of Maine

910

In Council, APR 7 1966

Department, Executive

ORDERED,

That the State Controller be authorized to make deductions from payrolls covering employees' membership in the American Federation of State, County and Municipal Employees - AFL-CIO, on the written authorization of each individual employee,

and

BE IT FURTHER ORDERED

That the State Controller pay over to the American Federation of State, County and Municipal Employees, AFL-CIO the amounts so deducted, ~OR ITS DESIGNEE

and

BE IT FURTHER ORDERED

That no such deductions shall be made unless and until at least 500 *valid* authorizations have been presented to the State Controller.

Statement of Fact

This Council Order is presented at the request of Mr. David Chisholm, International Representative of the SCME-AFL-CIO, which organization has been granted the privilege of dues deductions in the overwhelming majority of States.

A former Attorney General has ruled that similar payroll deductions are in order, on approval of the Governor and Council, if authorized by the individual employee.

Administrative Assistant

Read and passed by the Council, and by the Governor approved.

Deputy Secretary of State.

The forest commissioners have handled grievance cases with both unions following the prescribed five-step procedure for settlement. Most cases were settled and dropped at preliminary levels, a few were heard before the State Personnel Board, while only one went before the State Employees Appeal Board.

It should be noted that both the above labor organizations urged and supported the change of M.F.D. personnel from "unclassified" to "classified" service.

Appointments of the forest commissioners and of fire warden personnel have been particularly "clean" of any involvement in political patronage. There were a few isolated cases, but, as a whole there has been no interference in the efficiency of the M.F.D.'s program through "control appointments." The policy was and continues to be that only qualified, trained, and experienced personnel shall be employed regardless of party affiliation. A good illustration of this policy is found in a letter dated 1916, from which the following quote is taken: "I told him your only fault was in being a rank Democrat, and he did not care for that if you were a *good fire warden first*." Another illustration is found in a letter written by a former governor a number of years ago: "In general, it will be my purpose to have the Forestry Department managed on the basis of efficient service and kept as far away from politics as possible."

Records show no serious cases of violation of the federal Hatch Act against conflict of interest. This immunity from politics has contributed in a large measure to the fine spirit of cooperation with the governor's office, groups, associations, landowners, the Legislature, and the general public in regard to the program of forest protection.

Along with the continual change in employee relations, the historical record of the M.F.D. discloses an ever increasing emphasis upon training and communication. One of the most important aspects of this program was the institution of annual training meetings.

Forest Commissioner Forrest Colby is credited with initiating the first of these meetings between the timberland owners, the commissioner, Augusta office staff, and chief wardens. The year was 1919, and the purpose was to provide the chance for intercommunication and to assist in the appointment of fire wardens.

Annual notices of these meetings were sent out to all those concerned.* The meeting place alternated between Bangor, and Augusta. The chief wardens and landowners came to these meetings by train, with arrangements for meals and lodging made at a convenient hotel such as the Penobscot Exchange and Bangor House in Bangor or the Augusta House in Augusta.

Since those early days several changes have occurred in the type of landowner-warden meetings.

With the creation of the M.F.D.'s advisory committee in 1948, the Forestry Department has conducted its own annual forest warden school, with separate meetings between the commissioner and his

---

* See Appendix VII for samples of these notices and a program of the first meeting.

advisory committee. The training sessions dealt with all aspects of forest fire control, and the meetings concerned budget and policy matters of the M.F.D.

Later the training sessions became more general, covering all forestry related activities, which were, in turn, broken down into more specific sectional meetings. Under the heading of "fire control," annual sectional meetings were held at Rumford, Augusta, Greenville, Bangor, Princeton and Presque Isle. Throughout this entire period (1919–1972) a continuous relationship has been maintained between the landowners and the fire wardens. The format of the field-training sessions varied, with representatives from industry often participating, a factor which was most beneficial to the programs.

Quite in keeping with the subject of training is the consideration of safety. The M.F.D. has always been safety conscious, both in relation to the performance of regular duties and in the much more hazardous occupation of fighting fire. Recently greater emphasis has been placed on safety under the federal Occupational Safety Health Act (OSHA). In addition, industry has become extremely safety conscious and this attitude has had its effect on the programs of the M.F.D.

The tabulation below gives the M.F.D.'s accident rate for 1969–71.

Year	No. of Accidents	No. of Lost Time Accidents	No. of Days Lost	Severity Rate	Frequency Rate
1969	17	1	37	15.9	4.31
1970	25	3	31	13.4	12.92
1971	14	6	185	796.8	25.84

$$\text{Severity Rate} = \frac{\text{No. of days Lost} \times 1{,}000{,}000}{\text{No. of Hours of Exposure}}$$

$$\text{Frequency Rate} = \frac{\text{No. of disabling injuries} \times 1{,}000{,}000}{\text{No. of Hours of Exposure}}$$

The number of hours of exposure used was 232,176. There have been no fatalities due to fire fighting in the M.F.D. Only one fire warden was killed a few years ago when his jeep overturned on Squaw Mountain Trail.

Before turning to matters pertaining to public relations in terms of apparel, one activity involving the employees at the various levels of the M.F.D. organization should be mentioned again. In 1958, Forest Commissioner A. Nutting inaugurated the first forestry field day, which became an annual affair continuing to the present.

The main thrust of such field days was to recognize on different occasions the various aspects of forestry relating to fire protection

FIFTIETH ANNIVERSARY OF M.F.D. (1909), UNIVERSITY OF MAINE, ORONO, MAINE 1959
Twenty-five year service pins and certificates were issued to each man. *Front row, left to right:* Wilbur Pierce, patrolman; Edmund Brower, entomologist; Robert Nash, state entomologist; Robert Hutton, supervisor. *Back row, left to right:* Austin Wilkins, forest commissioner; Ralph Bagley, chief warden; Clarence Robers, patrolman; Everett Grant, chief warden; Emery Lyons, chief warden; Hutch McPheters, chief warden; Lawrence Lowell, waterman

utilization, management, logging, tree nursery programs, etc. The program of the first field day commemorated the first lookout station at Squaw Mountain and has already been cited (see pages 97–98).*

There has been a marked change in the style of warden badges over the sixty-odd years of the M.F.D.'s existence. In the early days these were simple, circular emblems, nickel plated and embossed with the appropriate titles of chief, patrolman, watchman, and deputy. Later, a more sophisticated type was designed in the shape of a shield, made of bronze and embossed with the State Seal, the title of the wearer (district warden or chief warden) and the words Maine Forest Service. These varied in size according to the warden's rank. A much smaller type was made for use on the cap, shirt, or coat.

---

* See Appendix VII for Foreword of printed program for the second field day, commemorating the 50th anniversary of the M.F.D.

A warden patch and samples of badges worn by members of the Maine Forest Service

Badges before the day of uniforms were important for identification as law enforcement officers.

Shoulder patches were designed in the form of a two-and-a-half- by three-and-a-half-inch patch in the shape of the State of Maine with "Maine Forest Service." This design was changed to a half-crescent in color for dress uniforms. Both designs had embroidery work.

As the M.F.D.'s forest fire wardens began to meet increasing numbers of people in the regular course of their public and law enforcement duties, the need for a uniform became very apparent. I made several legislative attempts as forest commissioner to gain funds for this purpose, and finally in 1960 an appropriation of $5,000 was made from the General Fund. This was an accomplishment, for some legislators, while feeling that members of the State Police and Inland Fisheries and Game wardens should be in uniform, seriously doubted the need in the case of forest fire wardens. After considerable persuasion and with the support of facts, the point was gained. It is of interest to bring out some of the reasons offered to support the case for uniforms. Forest fire wardens are law enforcement officers involved in long and irregular hours, subjected to a rigorous occupation that is often hazardous and constantly in the public eye, whether in issuing campfire and burning permits, enforcing the forest fire law, patrolling, conducting training sessions, or attending public meetings. In all these respects the forest fire warden differs in no way from police or game wardens. An added value of the uniform is evident when one remembers that the forest fire warden is often called to cooperate with other agencies during such emergencies as floods, hurricanes, and other natural disasters.

The end result of the adoption of a uniform for the personnel of the fire warden service was added stature, morale, dignity, and better law enforcement.

Once the appropriation was made available, a special departmental committee worked out the standard type of uniform with instructions for issuance of a suit, cruiser jacket, tie, shirt, hat, work pants, coveralls, caps, etc., to each warden and ranger. These are distributed free, with replacements when needed and justified. The use of such uniforms has now been expanded to include other branch members of the Forestry Department.

Earlier in this account of the M.F.D., mention was made of how a suggestion of a warden may have led to the "Keep Maine Green" slogans and to a possible "first" in the nation-wide program of alerting the public to their part in protecting our forests. The "Keep Maine Green" program started in 1948, but it was not until 1955 that a record was kept of the annual slogans.*

In 1942, Maine pioneered another program which had as its purpose the more precise measurement of forest conditions leading to an increased hazard of fire. This program also served in alerting the public to such dangers. The result of this experimentation and adoption of better techniques in cooperation with the U.S. Forest Service and the U.S. Weather Bureau was the creation of Fire Danger Stations. Initial stations were established within the M.F.D. and specifically in the St. John district. The system was expanded in later years to a state-wide network for a total of over forty-three stations. This number has been reduced; however, many stations were still operating as of 1973. Each station was located to cover an area of between 100,000 and 200,000 acres to form so-called "weather districts."

John Keetch of the U.S. Forest Service, now retired, spent considerable time in Maine perfecting certain aspects of the forest fire danger measurement. It was in Maine that another dimension was added to this system. In typical spruce-fir forests accumulations of

---

* See Appendix VII for list of "Keep Maine Green" slogans.

A most helpful forest fire prevention-education measure has been the use of forest fire danger class day signs and posters. Daily indicators are shown for the public to be aware of the forest fire danger days. Sizable roadside signs have been erected, others at fire stations, warden headquarters, also in radio and television stations, and for the press.

duff or humus range from a few inches to over two feet. It is these floor conditions that allow fires to burn deeply before reaching the underlying mineral soils. A factor for measuring the drying-out conditions of this "duff-humus" material has been worked out.

Forest fire danger measurements, a numerical scale ranging from "class one" to "class five" danger days, was developed with a range from low to extreme fire danger conditions. Such a system was a direct outgrowth of the continued study into relative fire occurrence and fire behavior. It became not only a most effective forest fire prevention tool, but also an effective educational device. Radio, television, and the press kept the public informed on the "class danger day." In addition large and small signs located at fire stations, warden headquarters, along roadsides, and at woods camps kept the public aware of conditions during the fire season.

Each morning the Forestry Department would broadcast over its radio network the weather forecast and the "class danger day." This service was also extended to the State Police, Civil Defense, fire departments, pulpwood camps, and landowners within the M.F.D. areas.

During the Centerville fire of 1963, which involved both Orga-

FIRE DANGER STATIONS
MAINE FORESTRY DISTRICT

District	Location	Attendant	Telephone	Weather Zone
NORTHERN DIVISION				
Allagash	Allagash Pl.	....	Dickey Central 2-13	3
Seven Islands	15 R 15 (St. Pamphile)	Harold Pelletier	St. Pamphile Central 27-12 Canadian Phone	3
Upper St. John	9 R. 18 (Hardwood Mt.)	Clifford Scott	Dickey Central 3-12 or 9-11	3
Chamberlain	8 R. 13 (Tramway)	F. Vaillancourt	Tramway through Greenville or Dickey Central 9-11	3
Aroostook Waters	Ashland	Harold A. Weeks	Ashland 3361	3
Aroostook Waters	11 R. 10--25 miles	Leslie G. Wakefield	Ashland 3361	3
Number 9	Hammond Pl.	Earl M. Adams	Houlton 4572	3
East Branch	6 R. 7 (Hay Lake)	Amy Davis	Patten 42-12	3
Katahdin	2 R. 9 (Togue Pond)	....	Millinocket 359-31	3
Mattawamkeag	Macwahoc Pl. (Whitney Hill)	Robert Graham	Sherman 9-13 or Wytopitlock 392	3
EASTERN DIVISION				
Pleasant River	6 R. 9 (Katahdin Iron Works)	....	Brownville 60	2
Passadumkeag	3 R. 1 (Sysladobsis)	....	Lee 1-21	2
St. Croix	Topsfield Twp. (Musquash)	Harry Noble	Princeton 10-2	2
East Machias	Cooper	William Dwelley	Meddybemps 657 M-11	2
Machias	Wesley (Main River)	Macey Armstrong	Columbia 35-2	2
WESTERN DIVISION				
Rangeley	Upton	....	Bethel 1-3	2
Rangeley	4 R. 2 (Cupsuptic)	Waylan Williams	Rangeley 80-13	2
Dead River	Eustis	Earle Williams	Stratton 16-4	2
Parlin Pond	Caratunk	Isac Harris	Bingham 2-1 or 3-2	2
Moose River	Jackman Pl.	Charles Lumbert	Jackman 107-3	2
Chesuncook	3 R. 12 (Chesuncook)	Oscar Gagnon	Greenville 50	3
Seboomook	2 R. 4 (Pittston Farm)	Vaughn Thornton	Greenville 50	3

nized Towns and the M.F.D., the Department for the first time had Monty Glovensky of the U.S. Weather Bureau in Boston establish a fire-weather danger station for purposes of hourly forecasts and predictions at the quarters of the fire boss.

Another program important to the public was the establishment of campsites. This program started in 1921, but only in the last decade was a budget item of $25,000 set up for intensifying this endeavor through the establishment of a network system of areas for the public to enjoy safe locations. Such a program, of course, was in the interest of forest fire prevention. The campsite coordinator, Temple Brown, did much to put this program on a firm operational basis. It is now functioning on a maintenance basis, with one hundred and seventy-five campsites handled by the regional directors. The largest number of campsites at any one time was three hundred and sixty, fifty-four of which were later turned over to the agency for state parks as being within the Allagash Wilderness Waterway.

It will be noted that there is an intimate relationship between forest fire prevention and public relations in all the M.F.D. programs mentioned in the preceeding paragraphs. This interrelationship is certainly evident in the final consideration to be offered in this chapter.

An effective forest fire prevention measure is the present statutory provision of woods closure by governor's proclamation. Just how many

Typical pulpwood camp, Appleton Township, Maine, 1914

fires have been prevented whenever this legislative act was invoked will never be known. However, its greatest value is the psychological impact upon the general public during periods of extreme woods drought conditions and serious on-going fires.

It is significant that the first woods closure legislation of Chapter 52, Sections 104, 1909, was enacted the same year as the creation of the Maine Forestry District. In subsequent years several amendments have been made for purposes of clarification and better law enforcement. Some of the changes were: closing all or sections of the state; annulment or lifting of the woods closure of the entire state or only certain sections; prohibiting smoking or building of out-of-door fires; suspension of the open seasons for fishing and hunting; and recommended closure action by the forest commissioner. Exceptions in the law permitted fishing from boats or canoes on lakes, ponds, rivers or thoroughfares, and hunting migratory waterfowl from boats in tidal water, or from offshore blinds.

In the administrative job of enforcement, the general public has always reacted with excellent cooperation and understanding. Similar cooperation has been received from fish and game wardens, fire wardens, woods operators and landowners.

In the entire history of this legislation there never has been a complete woods closure prohibiting anyone the right to travel or to earn a livelihood by working in the woods, provided he or she did not smoke, build fires, fish or hunt. During periods of extreme forest fire conditions many private woods operations have been temporarily closed down on a voluntary basis until conditions improved.

The closest approach to martial law under severe forest fire conditions occurred in the Bar Harbor fire of 1947 where most of the forest area is federally owned. There is no record of such action on privately owned or state lands.

In reviewing all the woods closure proclamations to date some interesting facts show up. Between the period 1909–1972, a span of 63 years, a total of thirty-one governors' proclamations have been issued. These occurred during the tenure of office of eleven governors and six forest commissioners.*

The first proclamation under the original act of 1909 was issued in 1911 by Governor Frederick W. Plaisted when Frank E. Mace was

---

* Woods Closure, Woods Ban and Fire Ban are not official terms and do not appear in the statute under Title 12, Section 1151, 1964 M.R.S.A. However, it is common terminology by the press and general public during dry periods and in governors' proclamations, and continues to be accepted usage. Tabulation of the sequence of the original closure law and amendments and all proclamations issued to date, together with samples of proclamations are included in Appendix VII.

forest commissioner. This was more of a warning than a prohibition of smoking or building out-of-door fires. In that year there was a state total of 202 fires burning over 111,077 acres, of which 127 fires, 99,654 acres and a damage of $389,052 occurred in the Maine Forestry District.

FIRST PROCLAMATION BY THE GOVERNOR OF MAINE REGARDING FOREST FIRES

WHEREAS, the towns, villages and timber lands of this State are in great danger from fire at the present time, owing to the almost unprecedented dry weather at this season of the year, and

WHERAS, our statutes contain the following provisions, wisely enacted for the protection of the lives and property of our people:

"Whoever by himself, or by his servant, agent or guide, or as the servant, agent or guide of any person, shall build a camp, cooking or other fire, or use an abandones camp, cooking or other fire, in or adjacent to any woods in this state, shall, before leaving such fire, totally extinguish the same, and upon failure to do so such person shall be punished by a fine of fifty dollars,– – –"

"Selectment shall erect in a conspicuous place at the side of every highway as they may deem proper, and at suitable distances alongside the rivers and lakes of the state frequented by camping parties, tourists, hunters and fishermen, in their respective towns, notices in large letters to be furnished by the Forest Commissioner, substantially in the following form: 'Camp fires must be totally extinguished before breaking camp, under penalty of not to exceed one month's imprisonment or one hundred dollars fine, or both as provided by law. Forest Commissioner.' The forest commissioner shall furnish owners of wood lands situated within this state when called upon so to do, notices of similar tenor to be posted at the expense of said owners upon their respective lands." – – – Sections 55 and 56, Chapter 7, R.S.

"Whoever kindles a fire on land not his own, without consent of the owner, forfeits ten dollars; if such fire spreads and damages the property of others, he forfeits not less than ten, nor more than five hundred dollars; and in either case he shall stand committed until fine and costs are paid, or he shall be imprisoned not more than three years."

"Whoever with intent to injure another, causes a fire

to be kindled on his own or another's land, whereby the property of any other person is injured or destroyed, shall be fined not less than twenty, nor more than one thousand dollars, or imprisoned not less than three months, nor more than three years."

"Whoever for a lawful purpose kindles a fire on his own land shall do so at a suitable time and in a careful and prudent manner; and is liable, in an action on the case, to any person injured by his failure to comply with this provision." Sections 15, 16 and 17, Chapter 28, R. S.

The State Forest Commissioner is using every possible means for the prevention of fires, and now has an organized force of three hundred men employed as wardens, lookout men and patrolmen, besides a large force of emergency men. In some sections of the state every available man is engaged in connection with this work and the large fires are under control, but smouldering, and likely to break out anew if vigilance is relaxed. The ponds, rivers and brooks are extremely low; we have had practically no rain since last October, and none whatever in the greater portion of the state for twenty-eight days. These conditions, with the usually hot weather, make the danger much greater than many of our people realize.

I therefore earnestly recommend that all persons, river drivers, railroad crews, sportsmen and guides in particular, use the utmost precaution. Farmers and others should not build brush fires, and in no case should fires be built on grass or timber lands, or on shores of lakes or streams, while the present conditions exist. It is of especial importance that municipal officers post notices as required by law, and take such other action as in their judgment will secure the co-operation of the citizens in every way that will tend to minimize the danger.

Given at the Executive Chamber, at Augusta, this twenty-second day of May, in the year of our Lord one thousand nine hundred and eleven, and of the Independence of the United States of America the one hundred and thirty-fifth.

Frederick W. Plaisted,
By the Governor.

In 1947 there was a reverse situation: the greatest number of fires, acres burned and losses occurred in the Organized Towns with the Maine Forestry District relatively free. The statistics for the state

show a total of 700 fires, 213,547 acres burned, and estimated damages of $11,990,855, of which only 167 fires, 4,685 acres burned, and damage of $20,164 were in the Maine Forestry District. Yet extremely dry conditions existed state-wide.

Governor Percival P. Baxter was the first and only governor to specifically make a reference to the *"Forestry District"* in his proclamation of 1921.

A summing up of the proclamations issued follows:

Most proclamations by a governor: six by Governor Horace A. Hildreth—1 in 1945; 1 in 1946; 2 in 1947; 2 in 1948.

Most proclamations recommended by a forest commissioner: nine by Commissioner Raymond E. Rendall, 1941–1947.

Longest period of closure for entire state: 35 days from July 28 to September 1, 1949.

Longest period of closure for entire state and certain sections extended: 39 days from July 25, 1946, and continued to August 28 for Hancock and Washington counties.

Shortest period of closure: three days, October 14 to 16, 1930.

Proclamations under various amendments:

1911, 1st proclamation—warning only

1922, 1st proclamation—suspension of open season for hunting

1930, 1st proclamation—suspension of open season for fishing

1938, 1st proclamation—closing only certain sections of the state

1945, 1st proclamation—lawful to build fires at M.F.D. authorized campsites

10 spring proclamations; 21 fall proclamations (1911–1972).

There are other examples under varying conditions.

Certain situations arose which necessitated special mention in the proclamations so as not to inconvenience the public. The date of issuance was usually not effective until the *next* day so as to give the public sufficient warning. These often read "effective at *midnight, sunrise, noon* or *sunset*." This gave some advance public notice to warn those already in the woods or those who were making plans to go camping, fishing or hunting.

The same notice applied to annulments. It is interesting to point out that during any extended period of closure the psychological effect soon begins to wear off, the public gets impatient, and the pressure is on for lifting the woods ban. The basis for recommending annulment is usually only after a general one-half inch to one inch of

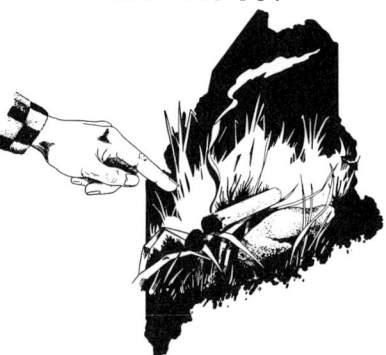

232 Industrial bilingual forest fire prevention signs put up in cooperation with the Maine Forest Service

PERMIS DE FEU EST EXIGÉ PAR LA LOI
FIRE PERMIT REQUIRED BY LAW

 Forest Commissioner Augusta, Maine — PREVENT FOREST FIRES

ATTENTION
EMPECHEZ LES FEUX DE FORETS

 FOREST COMMISSIONER AUGUSTA, MAINE

NE FUMEZ PAS EN MARCHANT DANS LES BOIS

DO NOT SMOKE WHILE WALKING IN THE WOODS

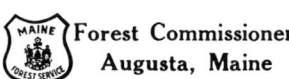

ATTENTION AU FEU LA FORÊT EST VOTRE GAGNE-PAIN

BE CAREFUL ABOUT FIRE THE FOREST IS YOUR MEANS OF LIVELIHOOD

rain. This brings danger station build-up index readings to zero and the drying out process of the forest floors, and days since last rain announcements start over again.

Any recommendation by the forest commissioner to the governor for a woods closure requires a careful appraisal of many factors. It means checking with fire wardens, landowners, woods operators, U.S. Weather Bureau, Inland Fisheries and Game Department, on the number of days since last appreciable rain, number and seriousness of on-going fires, and current fire danger station readings.

The Inland Fisheries and Game Department is the most likely to be affected by a woods closure since it is entirely dependent upon licenses and fees as its main source of income. A lengthy suspension of the open season for fishing and hunting usually results in a loss of income. On some occasions, it has been necessary for that Department to introduce a legislative bill requesting an increase in licenses and fees to meet these losses. In an off legislative year, funds were obtained from state surplus through governor and council action and reimbursed at the next regular session of the Legislature. Sporting camps have been similarly affected but are not able to recover their losses so easily. However, all recognize the need to protect the forests from fires.

Two freak acts of nature automatically brought about a woods closure: (1) the hurricane in the fall of 1938 caused an extremely high forest fire danger situation due to the millions of feet of wind-thrown timber and the resultant slash from salvage operations; and (2) the unusual three-day continuous high winds in October of 1947 fanned fires into an unbelievable fury and caused explosive woods conditions and heavy losses to timber and other property.

Considerable coverage has been given to this subject because of the direct relationship of the Maine Forestry District and the state as a whole brought about through any woods closure by governor's proclamation.*

CONCLUSION

In this Maine Forestry District story an attempt has been made to recapture and record for the first time a continuity of historical and interesting events of a remarkable era of forest protection. There is good reason to question if it will ever be equaled. "It has been a record of preparation and defense rather than disaster," as one landowner has so aptly stated.

Many still ponder the fact that the protection system worked so successfully. Much credit must go to those early and far-sighted forest owners who realized the need for protection of their holdings against fire and other natural enemies of the forest and took necessary action.

Unquestionably the forest protection program has greatly influenced and shaped the character and economy of the M.F.D. Today it has paid big dividends to landowners, the wood-using industry, and the general public.

Although not an exhaustive study, it is believed that in this history the major objective has been accomplished by updating and preserving the record of achievements. It would have been tragic to allow this to pass into oblivion.

With the passage of the Tree Growth Tax Law (Chapter 616, P.L. 1972) and creation of the Department of Conservation (Chapter 400, P.L. 1973) the curtain finally rang down on the Maine Forestry District as a separate entity. But what a distinguished record it has been of sixty-three consecutive years of forest protection in the unorganized territory of over ten million acres of timberland!

The end did not occur suddenly but rather as the result of gradual changes in economic conditions, which was inevitable. It was a movement from a single entity of a forest protection system to a state-wide administrative program to protect all the 18,000,000 acres of the forest resources of Maine.

The natural question most commonly asked is where do we go from here? By no means is this the end of the line. There must be continued effort in research, prevention-education, adequate funding, strengthening of laws and new advances in tactics and techniques toward the common goal of even greater reduction of acres burned and losses from timber and other values of the forest. To get the job done calls for accelerated cooperation between state and cooperating agencies, the private sector, and the general public.

Former Governor Kenneth M. Curtis made the following statement in one of his public addresses. "More demands are made of our forests—not only for wood fiber but recreation, wilderness areas, urbanization, water power, roads, utility lines, etc. But nature cannot by itself indefinitely provide without careful planning for adequate protection. In the long-term scale of things, people will continue to need our forests more than our forests will need people."

Finally, this historical documentary has been written with the hope that the reader will discover a deep sense of appreciation for the values of the forest and the contributions made by the protective systems of the Maine Forestry District and the good stewardship by forest owners of the land.

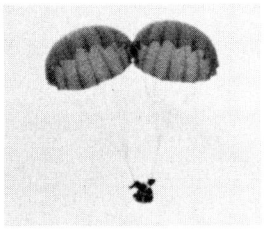

APPENDICES

I.	1. Maine Forestry District County Land Designations	239
	2. Conservation: Liability of Landowners	241
	3. Records of the Maine Land Office	242
	4. Table: Returns of Timber Cut in Unorganized Townships per County	244
	5. Table: Returns of Timber Cut in Unorganized Townships and Plantations	245
	6. Table: M.F.D. Assessments 1909–72	246
II.	1. M.F.D. Tax Legislation	248
	2. Table: Federal and State Allocations to Maine Under the Weeks and Clarke-McNary Laws	250
	3. Tables: M.F.D. Expenditures, 1917 and 1927	251
	4. Table: M.F.D. Budget and Operating Statement, 1950	252
	5. Table: M.F.D. Expenditures 1972	253
	6. Reasons for Significant Variances 1971 Budget vs. 1971 Actual Expenditures	254
III.	Foreword to Northeastern Forest Fire Protection Commission Manual	256
IV.	1. Table: Forest Fire Record	257
	2. Notes from Diary of Elmer B. Crowley	260
	3. M.F.D. Lookout Towers Still Operated in 1973	263
	4. M.F.D. Lookout Towers 1917	264
	5. Lookout Stations Operated in 1932	265
	6. M.F.D. Lookout Towers	266
	7. Cooperative Towers	268
V.	1. Table: Equipment Inventory	269
	2. Table: Capital Valuations	269
	3. Table: M.F.D. Suppression Costs 1917–72	270
	4. Samples of Forest Fire Reports	271
	5. An Affidavit	275
VI.	1. M.F.D. Radio Call Numbers, 1953 Directory	276
	2. Call and Car Plate Numbers	278
VII.	1. State Land Agents	279
	2. Artistic North Points from Early Documents	280
	3. Honorary Chief Fire Wardens	281
	4. Chief Wardens in Charge of Railroad Patrol	282
	5. Deputies Bangor & Aroostook Railroad	284
	6. Deputies Canadian Pacific Railroad Co.	285
	7. Deputies Maine Central Railroad	285
	8. Deputies Sandy River and Rangeley Lakes Railroad	286
	9. Wardens Appointed in 1909 with some Reminiscences	287
	10. Laws Regarding Appointments of Fire Wardens	288

11. State Employees Appeals Board   289
12. Chapter 147 re Unclassified State Forestry Dept. Employees   290
13. Salary Memo   291
14. Samples of Certificates of Appointment   292
15. Sample Announcements Timberland Owners
       Annual Meetings   296
16. Guardians of the Forests of Maine Program   297
17. Fiftieth Anniversary of the M.F.D.   299
18. Keep Maine Green Slogans   300
19. Original Closure Law   301
20. Summary of Woods Closure Proclamation   302
21. Sampling of Governors' Proclamations   304

Appendix I

1.   *Maine Forestry District—1972*
Approximately 10,483,867 Acres (Revised)

§ 1201. **Designation**

The administrative district known as the Maine Forestry District, heretofore established and incorporated, shall include the following territory.

Aroostook County. Townships W.E.L.S.: A,R.2; C,R.2; D,R.2; Cox Patent; 3,R.2; 3,R.3; 4,R.3; 7,R.3; 8,R.3; 9,R.3; 10,R.3; 17,R.3; 1,R.4; 2,R.4; 3,R.4; 7,R.4; 8,R.4; 9,R.4; 10,R.4; 11,R.4; 16,R.4; 17,R.4; A,R.5; 1,R.5; 7,R.5; 8,R.5; 9,R.5; 13,R.5; 14,R.5; 15,R.5; 16,R.5; 17,R.5; 10,R.6; 14,R.6; 15,R.6; 16,R.6; 9,R.7; 10,R.7; 11,R.7; 12,R.7; 13,R.7; 14,R.7; 9,R.8; 10,R.8; 11,R.8; 12,R.8; 13,R.8; 14,R.8; 15,R.8; 16,R.8; 11,R.9; 12,R.9; 13,R.9; 14,R.9; 15,R.9; 16,R.9; 11,R.10; 12,R.10; 13,R.10; 14,R.10; 15,R.10; 18,R.10; 11,R.11; 12,R.11; 13,R.11; 14,R.11; 15,R.11; 18,R.11; 19,R.11; 11,R.12; 12,R.12; 13,R.12; 14,R.12; 15,R.12; 16,R.12; 17,R.12; 18,R.12; 19,R.12; 20,R.11 and 12; 11,R.13; 12,R.13; 13,R.13; 14,R.13; 15,R.13; 16,R.13; 17,R.13; 18,R.13; 11,R.14; 12,R.14; 13,R.14; 14,R.14; 15,R.14; 16,R.14; 17,R.14; 11,R.15; 12,R.15; 13,R.15; 14,R.15; 15,R.15; 11,R.16; 12,R.16; 13,R.16; 14,R.16; 11,R.17; 12,R.17; Silver Ridge. Municipalities: Allagash Plantation, E Plantation, Garfield Plantation, Glenwood Plantation, Hammond Plantation, Macwahoc Plantation, Nashville Plantation, Oxbow Plantation, Reed Plantation, Wallagrass Plantation, Westmanland Plantation, Winterville Plantation.

Franklin County. Townships B.K.P.; W.K.R.; 4,R.1; 3,R.2; 4,R.2; 4,R.3. Townships W.B.K.P.: 1,R.2; 2,R.3; 3,R.3; 2,R.4; 3,R.4; 1,R.5; 2,R.5; 3,R.5; 1,R.6; 2,R.6; 3,R.6; Gore N. 2 and 3,R.6; 1,R.7; 2,R.7; 1,R.8; Gore N. 1,R.8; 2,R.8. Other townships: D; E; 6,N. of Weld; Perkins; Washington; Freeman; Salem. Municipalities: Coplin Plantation, Dallas Plantation, Rangeley Plantation, Sandy River Plantation.

Hancock County. Townships N.D.: 3 and Strip North; 4 and Strip North. Townships S.D.: 7, 8, 9, 10. Townships M.D.: 12, 22, 28, 32, 34, 35, 39, 40, 41. Islands: Beach, Bear, Bradbury's Butter or Dirigo, Eagle, Hog, Little Spruce, Marshall's, Pickering's Pond, Resolution, Spruce Head, Western. Municipalities: Osborn Plantation, No. 33 Plantation.

Oxford County. Townships W.B.K.P.: 4,R.1; 4,R.2; 4,R.3; 5,R.3; 4,R.4; 5,R.4; 4,R.5; 5,R.5; 4,R.6. Other townships: A,1 (Riley); A,2 (Grafton); Andover North Surplus; Andover West Surplus; C; C Surplus; Albany; Mason. Municipalities: Lincoln Plantation, Magalloway Plantation.

Penobscot County. Townships N.B.P.P.: 3,R.1; 5,R.1. Townships N.W.P.: 1,R.7; 2,R.8; 2,R.9; 3,R.9; Townships W.E.L.S.: 1,R.6; 2,R.6; 6,R.6; 7,R.6; 8,R.6; A,R.7; 1,R.7; 2,R.7; 3,R.7; 4,R.7; 5,R.7; 6,R.7; 7,R.7; 8,R.7; 1,R.8; 2,R.8; 3,R.8; 4,R.8; 5,R.8; 6,R.8; 7,R.8; 8,R.8; Hopkins academy Grant; A,R.8 and 9; Veazie Gore. Other townships: 3 Indian Purchase, 4 Indian Purchase; 1, N.D.; Kingman. Municipalities: Medway, Drew Plantation, Grand Falls Plantation, Lakeville Plantation, Seboeis Plantation, Webster Plantation.

Piscataquis County. Townships N.W.P.: 6,R.8; 4,R.9; 5,R.9; 6,R.9; 7,R.9; 7,R.10; 8,R.10; Townships B.K.P., E.K.R.: 3,R.5; 2,R.6. Townships W.E.L.S.: 1,R.9; 2,R.9; 3,R.9; 4,R.9; 5,R.9; 6,R.9; 7,R.9; 8,R.9; 9,R.9; 10,R.9; A,R.10; B,R.10; 1,R.10; 2,R.10; 3,R.10; 4,R.10; 5,R.10; 6,R.10; 7,R.10; 8,R.10; 9,R.10; 10,R.10; A,R.11; B,R.11; 1,R.11; 2,R.11; 3,R.11; 4,R.11; 5,R.11; 6,R.11; 7,R.11; 8,R.11; 9,R.11; 10,R.11; A,R.12; 1,R.12; 2,R.12; 3,R.12; 4,R.12; 5,R.12; 6,R.12; 7,R.12; 8,R.12; 9,R.12; 10,R.12; A,R.13; A2,R.13 and 14; 1,R.13; 2,R.13; 3,R.13; 4,R.13; 5,R.13; 6,R.13; 7,R.13; 8,R.13; 9,R.13; 10,R.13; A.R.14; 1,R.14; 2,R.14; 3,R.14; 4,R.14; 5,R.14; 6,R.14; 7,R.14; 8,R.14; 9,R.14; 10,R.14; East Middlesex Canal; Days Academy Grant; 3,R.15; 4,R.15; 5,R.15; 6,R.15; 7,R.15; 8,R.15; 9,R.15; 10,R.15. Other townships: Harford's Point; Cove Point; All islands in Moosehead; Medford; Orneville. Municipalities: Bowerbank; Barnard Plantation; Elliotsville Plantation; Kingsbury Plantation, Lakeview Plantation.

Somerset County. Townships B.K.P., W.K.R.: 2,R.1; 1,R.3; 2,R.3; 3,R.3; 4,R.3; 1,R.4; 2,R.4; 3,R.4; 4,R.4; 2,R.5; 3,R.5; 4,R.5; 1,R.6; 2,R.6; 3,R.6; 4,R.6; 5,R.6; 1,R.7; 2,R.7; 3,R.7; Gore N. of T.1,2 and 3,R.7; 4,R.7; 5,R.7; 6,R.7; Townships B.K.P.; E.K.R.: 2,R.2; 2,R.3; 2,R.4; 1,R.5; 2,R.5; 1,R.6. Townships N.B.K.P.: 1,R.1 (Tauton and Raynham); Sand Bar Tract; 1,R.1 (Rockwood Strip); 2,R.1 (Sandwich Academy Grant); 2,R.1, (Rockwood Strip); 3,R.1; 5,R.1; 6,R.1; 1,R.2; 2,R.2; 3,R.2; 6,R.2; Big W; Little W; 1,R.3; 2,R.3; 3,R.3; 4,R.3; 5,R.3; Seboomook, R.4; 1,R.4; 2,R.4; 3,R.4; 4,R.4; 5,R.4; 3,R.5; 4,R.5. Townships W.E.L.S.: 4,R.16; 5,R.16; 6,R.16; 7,R.16; 8,R.16; 9,R.16; 10,R.16; 4,R.17; 5,R.17; 6,R.17; 7,R.17; 8,R.17; 9,R.17; 10,R.17; 4,R.18; 5,R.18; 6,R.18; 7,R.18; 8,R.18; 9,R.18; 5,R.19; 6,R.19; 7,R.19; 8,R.19; 5,R.20. Other townships: Concord. Municipalities: Moose River, Moscow, Brighton Plantation, Caratunk Plantation, Dennistown Plantation, Highland Plantation, Pleasant Ridge Plantation, The Forks Plantation, West Forks Plantation.

Washington County. Townships E.D.: 18; 19; 26; 27. Townships M.D.: 18; 19; 24; 25; 29; 30; 31; 36; 37; 42, 43. Townships N.D.: 5, and strip North; 6, and strip North. Townships T.S.: 1,R.1; 1,R.2; 1,R.3. Townships N.B.P.P.: 6,R.1; 7,R.2; 8,R.3; 10,R.3; 11,R.3; 8,R.4; 9,R.4. Other townships: Brookton, Edmunds, Indian, Marion, Trescott. Municipalities: Beddington, Centerville, Cooper, Crawford, Deblois, Northfield, Topsfield, Wesley, Baring Plantation, Codyville Plantation, Grand Lake Stream Plantation, No. 14 Plantation, No. 21 Plantation.

Whenever it shall appear to the State Tax Assessor that any part of the unorganized territory of the State, including any areas previously incorporated but which have been deorganized by Act of Legislature, is in need of fire protection, the State Tax Assessor with the approval of the Forest Commissioner and the Attorney General may declare such territory to be part of the Maine Forestry District.

§ 1202. Adjacent municipalities as part of district

Any municipality adjoining any part of the Maine Forestry District may, by vote at any meeting of its inhabitants duly called and held, become a part of said Forestry District and subject to all the provisions of this subchapter, and subchapter IV and X. A copy of such vote, certified by the municipal clerk shall be forwarded forthwith to the State Tax Assessor, to the Treasurer of State and to the commissioner, and from the time such certified copy is filed in the office of the Treasurer of State, the municipality so voting shall be and continue a part of said Forestry District. All municipalities which shall become a part of said district and all officers of such municipalities shall be and are exempt from the duties and obligations imposed by Title 25, chapter 319.

TITLE 12: CONSERVATION

CHAPTER 333: LIABILITY OF LANDOWNERS

2. § 3001. Definitions

The word "premises" as used in this chapter includes lands, private ways and any buildings and structures thereon.

§ 3002. No duty to keep premises safe or give warning

An owner, lessee or occupant of premises owes no duty to keep the premises safe for entry or use by others for hunting, fishing, trapping, camping, hiking, sight-seeing, operation of snow traveling vehicles or recreational activities, or to give warning of any hazardous condition or use of or structure or activity on such premises to persons entering for such purposes, except as provided in section 3004.

§ 3003. Permission as affecting liability

An owner, lessee or occupant of premises who gives permission to another to hunt, fish, trap, camp, hike, sight-see, operate a snow traveling vehicle or participate in recreational activities upon such premises does not thereby extend any assurance that the premises are safe for such purpose, or constitute the person to whom permission is granted an invitee to whom a duty of care is owed, or assume responsibility for or incur liability for any injury to person or property caused by any act of persons to whom the permission is granted, except as provided in section 3004.

§ 3004. Liability

This chapter does not limit the liability which would otherwise exist for willful or malicious failure to guard, or to warn against, a dangerous condition, use, structure or activity; or for injury suffered in any case where permission to hunt, fish, trap, camp, hike, sight-see, operate a snow traveling vehicle or participate in recreational activities was granted for a consideration other than the consideration, if any, paid to said landowner by the State; or for injury caused by acts of persons to whom permission to hunt, fish, trap, camp, hike, sight-see, operate a snow traveling vehicle or participate in recreational activities was granted, to other persons as to whom the person granting permission, or the owner, lessee or occupant of the premises, owed a duty to keep the premises safe or to warn of danger.

§ 3005. **No duty created by statute**

Nothing in this chapter creates a duty of care or ground of liability for injury to person or property.

Note: Enacted 1961 Chapter 296

Attorney Thomas Weeks, of Waterville, Maine, was largely responsible for drafting much of the material for this piece of legislation.

---

3. RECORDS OF THE MAINE LAND OFFICE

The following selected list summarizes the principal series contained in the records of the Maine Land Office:

DEEDS AND RELATED MATERIALS, 1794–1949

    Massachusetts Deeds, 1794–1828.
    Records of Deeds of Confirmation, 1841–1843.
    Treaty Deeds, 1868–1879 (Maine lands confirmed under the Webster-Ashburton Treaty, 1842).
    Record of Deeds, 1828–1940.
    Deeds of Land Sold for Taxes, 1848–1854; 1909–1945.
    Records of Land and Settlers' Certificates, 1842–1884.
    "Miscellaneous" Records of Deeds, 1798–1949.

FIELD NOTES, 1803–1890

The field notes give information about boundary lines, forest growth, topography, distances, and related information about the areas surveyed. They frequently contain information about hardships or unusual occurrences encountered during the survey.

MAPS

The maps prepared by the Massachusetts and Maine Land Offices, or accumulated by those agencies and the Maine Forestry Department represent various surveys and lottings of boundary lines undertaken since the eighteenth century. From time to time, the Maine Legislature authorized funds for the copying of early maps of Maine held by Massachusetts for the benefit of the Land Office. Other maps included in this series were prepared by various private concerns and acquired by the Forestry Department. A comprehensive index to the maps is arranged by the various counties of the State; and thereunder by town, township or plantation.

OTHER LAND OFFICE AND FORESTRY DEPARTMENT RECORDS

The records of the Maine Land Office and related materials generated by the Forestry Department were maintained, indexed and controlled by those agencies in such a way as to facilitate the location of specific information as it was needed by surveyors and those engaged in title research. Such use continues at a high volume, and the Maine State Archives has had to retain the Land Office and Forestry Department filing order and accompanying indexes to meet the needs of surveyors and title searchers at the expense of traditional archival control procedures. Many other records of the Land Office that are not included in the above list bear, nonetheless, a close relationship with the primary records series of maps, field notes and deeds; but cannot be treated as component parts thereof. Still other documentation that emanated from field offices of the various Land Agents, or that was created by special surveys instituted by the Legislature, or produced by the efforts of private corporations or individuals was incorporated in the mass of the Land Office records and is now difficult to isolate and identify. A further complication has resulted from the former practice of binding into one or more volumes copies of records relating to a particular subject or to a specific activity undertaken over a short period of time. These can only be identified by means of an item-by-item listing, a practice unsuitable for a general descriptive brochure of this type. Among these diverse records are the following, which may be of particular interest to the researcher:

- RECORDS, PROCEEDINGS AND SURVEYS UNDERTAKEN BY THE BOARD OF COMMISSIONERS UNDER THE ACT OF SEPARATION
- RECORDS RELATING TO THE ACTIVITIES OF THE BOARD OF COMMISSIONERS FOR THE SALE OF EASTERN LANDS
- SURVEYS UNDERTAKEN UNDER THE WEBSTER-ASHBURTON TREATY, 1842
- SCHEDULES OF LAND SALES AND RELATED CERTIFICATES
- MISCELLANEOUS JOURNALS, MEMORANDA BOOKS AND CORRESPONDENCE
- ROAD LABOR NOTES AND ROAD LABOR AGENTS' ANNUAL RETURNS
- REPORTS, INVENTORIES AND OTHER RECORDS RELATING TO THE PUBLIC LOTS IN ORGANIZED AND UNORGANIZED TOWNSHIPS
- APPLICATIONS AND CERTIFICATES FOR BOUNTY LANDS AND MONEY (Veterans of the American Revolution and War of 1812)

Researchers may wish to consult the records of the State Bureau of Taxation and the State Board of Tax Assessors for supportive information related to the wildlands and the public domain. The various State environmental agencies can provide recent and updated information about the public lands of the State.

Prepared by Maine State Archives

4. RETURN OF TIMBER CUT IN UNORGANIZED TOWNSHIPS PER COUNTY
(From Biennial Bureau of Taxation Report - 1971-72)
July 1, 1970 - June 30, 1971

M. Feet Board Measure

Counties	Spruce & Fir	Cedar	White* Pine	Hemlock	Yellow Birch	Maple	Beech	Misc.
Aroostook	125,747	8,152	6,303	222	38,262	14,781	1,451	2,057
Franklin	6,023	24	356	198	796	185	12	1,211
Hancock	250	6	2,224	531	21	110	186	39
Kennebec			208					16
Oxford	3,878		285	74	607	385	10	202
Penobscot	2,754	4	6,258	174	572	2,750	394	2,252
Piscataquis	61,206	1,336	13,005	322	2,900	8,259	118	1,332
Somerset	42,170	1,406	2,551	69	3,717	9,144	24	2,428
Washington	660		1,396	265	59	90	5	851
TOTAL	242,688	10,928	32,586	1,788	46,934	35,704	2,200	10,388

Grand Total: 383,216 M Bd. ft.

CORDS
Pulpwood (Rough Basis) - Cords

Counties	Spruce & Fir	Hemlock	Misc. Softwood	Poplar	Hard-wood	Hwd. Fuel	White Birch	Bolts	Cedar Poles
Aroostook	332,084	38,418	2,145	365	26,748	165	5,791	26	1,979
Franklin	6,475	1,696	126	117	12,689		1,767	2,110	
Hancock	14,684	4,053	34	1,316	6,396	7	741		
Kennebec	6				49				
Oxford	13,429	770	203		20,319		393	159	
Penobscot	113,591	30,123	6,814	3,703	23,864	717	2,300	472	
Piscataquis	255,271	4,191	411	375	9,665	57	387	67	
Somerset	180,457	2,207	247	5	8,980	32	1,745	194	2
Washington	58,012	27,666	4,941	44	41,252	70	70	1,051	
TOTAL	974,003	109,130	14,921	5,925	149,962	1,048	13,194	4,079	1,981

Grand Total: 1,272,035 Cds.

*Includes 1.9 per cent Red Pine

5. RETURNS OF TIMBER CUT IN UNORGANIZED TOWNSHIPS & PLANTATIONS
Aroostook, Franklin, Hancock, Kennebec, Oxford, Penobscot,
Piscataquis, Somerset & Washington Counties
(Statistics from Bureau of Taxation Biennial Reports)

YEARS Fiscal year July 1-June 30	THOUSANDS OF BOARD FEET (Lumber)	Cords (Pulpwood)	PIECES		Ship Knees
			(RR) Ties	CEDAR Poles	
1920	359,337	530,065*	554,094	22,168	43
1921	205,933	989,868*	379,744	14,422	--
1922	192,848	447,153*	344,470	33,121	--
1923	271,501	768,707*	115,360	11,247	--
1924	234,257	641,663*	394,500	20,487	--
1925	169,287	437,203	281,172	11,028	--
1926	148,435	467,092	205,098	16,145	--
1927	136,364	641,071	159,686	14,557	87
1928	100,000	599,771	303,307	9,547	--
1929	126,804	530,833	426,744	12,731	45
1930	101,012	629,665	237,469	17,562	--
1931	62,792	548,681	176,184	52,771	--
1932	24,641	294,169	162,241	4,053	--
1933	61,078	194,387	124,159	2,184	--
1934	47,638	306,712	39,636	8,234	--
1935	91,115	563,913	115,018	2,815	--
1936	67,013	437,176	174,185	8,119	--
1937	81,815	456,694	67,465	14,524	--
1938	93,618	612,619	149,228	27,553	--
1939	59,390	259,641	147,120	44,004	--
1940	68,441	501,304	108,895	46,120	--
1941	64,026	503,363	81,194	31,598	--
1942	93,376	598,952	49,191	28,420	--
1943	95,739	667,406	8,235	12,758	--
1944	99,510	385,563	12,680	14,824	--
1945	100,947	474,020	5,610	13,486	--
1946	68,312	673,152	10,469	10,699	--
1947	104,986	585,430	16,617	8,473	--
1948	141,542	624,783	52,319	15,274	--
1949	134,149	567,426	45,639	18,538	--
1950	94,638	408,155	67,306	16,043	--
1951	129,084	563,513	11,576	4,863	--
1952	103,808	736,031	6,521	1,474	--
1953	181,347	634,752	26,558	4,381	--
1954	149,449	574,889	13,562	3,547	--
1955	130,874	578,458	9,031	1,433	--
1956	201,975	1,110,919	37,373	80,881	--
1957	208,060	1,089,883	34,508	11,970	--
1958	144,926	902,754	49,310	15,103	--
1959	181,472	601,628	21,962	7,951	--
1960	261,915	824,684	4,759	3,390	--
1961	273,881	1,164,167	3,024	12,849	--
1962	284,603	747,650	6,320	16,625	--
1963	233,351	817,686	4,308	9,701	--
1964	323,892	922,036	1,571	5,701	--
1965	365,706	1,050,364	32	9,410	--
1966	348,824	807,644	-----	10,503	--
1967	358,318	1,099,833	1,335	6,685	--
1968	363,106	1,018,863	19	3,852	--
1969	490,131	1,364,660	31	1,647	--
1970	468,737	1,266,016	750	1,298	--
1971	383,216	1,272,035	--	1,981	--
Totals (52 Yrs)	9,287,219	35,495,119	5,247,585	778,950	175

No more returns on this basis due to Tree Growth Law (Chapter 616, P.L. 1972). Annual timber cut tables have come from the Forestry Dept. on a Calendar Year basis since 1972.

*Small quantities of lumber and pulpwood were shown for Kennebec County 1920-1924 Incl., and after 1932 no timber was cut in this county. Lumber includes: white pine, spruce-fir, cedar, hemlock, yellow birch, poplar, beech, and miscellaneous hardwood. Pulpwood includes: spruce-fir, pine, hemlock, poplar and hardwood.

6. MAINE FORESTRY DISTRICT ASSESSMENTS
AS TAKEN FROM THE BIENNIAL REPORTS

YEAR	VALUATION	MILL RATE	AMOUNT
1909	$ 42,630,293.	1 1/2	$ 63,945.44
1910	42,630,293.	1 1/2	63,945.44
1911	45,281,647.	1 1/2	67,922.47
1912	45,281,647.	1 1/2	67,922.47
1913	46,938,873.	1 1/2	70,408.31
1914	47,444,207.	1 1/2	71,166.31
1915	48,886,053.	1 1/2	73,329.08
1916	48,890,693.	1 1/2	73,336.04
1917	55,290,473.	1 1/2	82,935.71
1918	55,503,813.	1 1/2	83,255.72
1919	64,442,211.	1 3/4	112,773.87
1920	64,977,954.	1 3/4	113,711.42
1921	69,797,138.	2 1/4	157,043.56
1922	70,177,031.	2 1/4	157,898.32
1923	73,331,973.	2 1/4	164,996.94
1924	72,427,978.	2 1/4	162,962.95
1925	73,587,675.	2 1/4	165,572.27
1926	73,561,520.	2 1/4	165,513.42
1927	74,434,298.	2 1/4	167,477.17
1928	74,434,298.	2 1/4	167,477.17
1929	75,104,124.	2 1/4	168,984.28
1930	75,104,124.	2 1/4	168,984.28
1931	80,595,311.	2 1/4	181,339.45
1932	80,620,880.	2 1/4	181,396.98
1933	70,612,542.	2 1/4	158,878.22
1934	70,612,542.	2 1/4	158,878.22
1935	60,950,555.	2 1/4	137,138.75
1936	61,129,933.	2 1/4	137,542.35
1937	59,333,098.	2 1/4	133,499.47
1938	59,501,982.	2 1/4	133,879.46
1939	59,385,169.	2 1/4	133,616.63
1940	59,399,382.	2 1/4	133,648.61
1941	59,206,947.	2 1/4	133,215.63
1942	59,221,271.	2 1/4	133,247.86
1943	59,341,995.	2 1/4	133,519.49
1944	59,552,880.	2 1/4	133,993.98
1945	59,624,667.	2 1/4	134,155.50
1946	59,657,947.	2 1/4	134,230.38
1947	59,818,733.	2 1/4	133,592.15
1948	59,686,893.	2 1/4	134,295.51
1949	60,539,980.	8	484,319.84
1950	60,657,804.	8	485,262.43
1951	63,365,288.	5 1/2	348,509.08
1952	63,366,569.	5 1/2	348,516.13
1953	70,439,019.	9 1/2	669,170.68
1954	70,441,496.	5 1/2	387,428.23
1955	97,493,832.	4 3/4	463,095.70
1956	97,493,832.	4 3/4	463,095.70

1957	$103,493,264	4 3/4	$ 491,593.00
1958	99,387,473	4 3/4	472,090.50
1959	105,184,851	4 3/4	499,628.04
1960	105,008,444	4 3/4	498,790.11
1961	111,348,935	4 3/4	528,907.44
1962	111,355,110	4 3/4	528,936.77
1963	110,326,876	4 3/4	524,052.66
1964	109,440,619	4 3/4	519,842.94
1965	113,056,789	5 1/4	593,548.14
1966	112,943,158	5 1/4	592,951.58
1967	113,465,335	9	1,021,188.02
1968	113,477,469	8	907,819.75
1969	126,475,942	8 1/2	1,075,045.51
1970	127,471,846	8 1/2	1,083,510.69
1971	157,736,629	8 1/2	1,340,761.34
1972	156,607,266	8 1/2	1,331,161.69

Total from 1909 $20,810,857.25

The above figures do not include any of the M.F.D. six periodic Spruce Budworm tax assessments which accumulatively amount to $912,220.78 (years of legislative authorization 1937-'59-'63-'67-'69-'71).

Often the public loses sight of the fact that landowners in the unorganized townships pay other taxes. With the exception of the Spruce Budworm tax assessment the full schedule of annual tax rates in the unorganized towns is shown below:

Tax Rates - Unorganized Townships

The overall net tax rate on real estate in any unorganized township can be determined by adding the component rates, as shown below. For example, the total net rate in Silver Ridge Township, Aroostook County, is .091518 mills, or $91.51 per $1,000 of valuation, as follows:

State Tax	(credited towards school capital and roads)
Forest District Tax	.01125
County Tax	.002228
School Tax	.03410
School Capital Tax	.015
Road Tax	.025
Fire Protection	.00394
	.091518 (or 91.518 mills)

State, Forestry District, or Forest Fire and County Taxes apply to all townships; road, school, school capital, fire protection, and public services apply only to certain townships as shown.

State of Maine
Bureau of Taxation
June, 1972
Revised July 26, 1972

Appendix II

1. M.F.D. Tax Legislation

§ 513. Authority to accept federal, municipal and private funds

The department is designated the public agency of the State for the purpose of accepting federal, municipal and private funds in relation to forest fire protection, insect and disease control, management, growth, research and related forest matters, excepting federal funds received under the Stennis-McIntire bill, Public Law 87–788. The Treasurer of State shall be the appropriate fiscal officer to receive such funds for these purposes, subject to the approval of the Governor and Council, and the State Comptroller shall authorize expenditures therefrom as approved by the department and the Governor and Council.

§ 1141. Taxation by State; forest fire tax

Real estate not exempt, and not liable to be assessed in any town, may be taxed by the Legislature for a just proportion of all state, county and forestry district taxes for ordering the state, county and forestry district taxes upon property liable to be assessed in towns. The State Tax Assessor shall make lists thereof, with as many divisions as will secure equitable taxation, conforming as near as convenient to known divisions and separate ownership.

All areas not incorporated outside the Maine Forestry District shall pay a forest fire tax equal to that of the Maine Forestry District. The valuation as determined by the State Tax Assessor and set forth in the statement filed by it as provided by section 381 or section 381-A shall be the basis for the computation and apportionment of the tax assessed. The sum of $50 of the amount assessed for each area shall be credited to the general forestry appropriation, forest fire control for organized towns, to allow the Forest Commissioner to employ a forest fire warden for prevention and the remainder credited to the aid to towns appropriation for control and suppression of forest fires.

§ 1142. Determination of tax; list filed for public inspection

When the real estate mentioned in section 1141 is assessed for any state, county and forestry district taxes, the State Tax Assessor shall determine the proportionate amount of such taxes due from the owners of such real estate by applying the total millage rate of all such taxes against the valuation as listed by the State Tax Assessors. The statements of the total tax due from each such owner shall be mailed as provided in section 1145. The State Tax Assessor shall make a list, using the last state valuation as established by him. Such list shall contain the total amount of any state, county and forestry district taxes due from each owner of real estate mentioned in section 1141 and each owner of rights in public reserved lots, and shall contain the millage rate used in determining the proportionate amount of taxes due from such owners. Such list shall be filed in the office of the State Tax Assessor on or before the first day of July of each year, and shall be available for public inspection.

§ 1601. Annual district tax

A tax of 9 mills on the dollar is assessed for the year 1967 and 8½ mills thereafter upon all the property in the Maine Forestry District, including rights in public reserved lots, to be used for the protection thereof; except that in organized municipalities the tax rate shall be 9 mills for the year 1967 and 8½ mills thereafter multiplied by a fraction whose numerator is the previous year's assessed value of the land taxable by the municipality, including dams and power houses but not including any other structure or building, and whose denominator is the total previous year's assessed value of all property taxable by the municipality. Such tax shall be increased by ½ mill on the dollar assessed only for the year 1967 upon all the property in the unorganized territory located within the Maine Forestry District, including rights in public reserved lots, to be used by the Forestry Department for spruce budworm control. Said tax shall be paid on or before the first day of October, annually. The valuation as determined by the State Tax Assessor, and set forth in the statement filed by him as provided by Title 36, sections 381 or 381-A, shall be the basis for the computation and apportionment of the tax assessed. The State Tax Assessor shall determine, in accordance with Title 36, section 1142, the amount of such taxes due from the owners of lands in each unorganized township and lot or parcel of land not included in any township and public reserved lots, and such amounts shall be included in the statements referred to in Title 36, section 1145. The tax assessed shall be included in the statements referred to in Title 36, section 1145. The tax assessed shall be valid, and all remedies provided shall be in full force if said property is described with reasonable accuracy, whether the ownership thereof is correctly stated or not.

§ 1601-A. Reimbursement to municipalities

The Maine Forestry District shall reimburse the member municipalities for costs incurred each year for fire protection other than what the Maine Forestry District provides, up to a maximum of 50% of the contribution of the respective municipality to the Maine Forestry District in that year. The amounts to be reimbursed hereunder shall be certified by the Forest Commissioner, which amounts are hereby appropriated to pay the same, and the Governor and Council may authorize the State Controller to draw his warrant therefor at any time. Said amounts shall be charged against the fund provided in section 1607.

§ 1606. Assessments on plantations

The Treasurer of State shall annually send his warrant, together with a copy of the assessment of taxes upon the plantations in the Maine Forestry District, directed to the municipal officers of said plantations, requiring them respectively to assess, in dollars and cents, the sum so charged according to the law for the assessment of such taxes, and to add to the amount of such tax the amount of state, county and plantation taxes to be assessed by them in each plantation respectively.

§ 1607. Use of funds; deficiency; payments from State Treasurer; audit

The tax assessed by authority of section 1601 shall be recorded on the books of the State in a separate account as a fund to be used to protect from fire the forests situated within the Maine Forestry District, and to pay expenses incidental thereto, including payment of wages of clerks in the department's offices, and for no other purpose. If the tax assessed by authority of section 1601 for any reason is not available for the purpose aforesaid or if said taxes prove insufficient in any year to properly carry out said purposes, the Governor and Council may make available for said purposes, from any moneys then in the treasury not otherwise appropriated, such sum or sums of money as they may deem necessary for such purposes. Except as provided, the expenditures of forestry district funds shall be in accordance with Title 5, chapters 7, 11, 141 to 155. Said chapters shall not otherwise apply to said Forestry District.

2. Federal and State Allocations to Maine under the Weeks (1911-1923) & Clarke-McNary (1924 on) Laws for Organized Towns and M.F.D.

Year	Federal	State	Year	Federal	State
1911	$ 9,986	$ 49,661	1946	$ 204,104	$ 204,105
1912	6,508	31,376	1947	167,743	370,815
1913	8,115	67,332	1948	137,848	531,180
1914	8,911	84,560	1949	154,815	579,187
1915	5,104	57,531	1950	170,210	670,483
1916	6,038	50,432	1951	162,881	773,043
1917	6,143	78,960	1952	291,451	900,762
1918	7,506	110,734	1953	197,974	879,200
1919	5,727	143,070	1954	188,740	567,541
1920	7,165	147,710	1955	204,187	690,766
1921	24,954	214,830	1956	160,510	660,372
1922	37,828	198,685	1957	321,124	787,807
1923	23,416	149,255	1958	213,825	757,048
1924	22,462	181,628	1959	170,825	879,856
1925	22,566	162,672	1960	233,400	1,160,765
1926	27,245	150,685	1961	359,500	816,310
1927	35,147	156,236	1962	379,700	985,428
1928	63,296	101,110	1963	308,700	1,006,652
1929	53,365	147,612	1964	294,315	1,112,179
1930	52,965	160,822	1965	360,720	1,487,657
1931	54,322	168,833	1966	323,240	1,331,319
1932	56,922	169,212	1967	382,172	1,270,184
1933	37,714	125,173	1968	352,998	1,586,385
1934	64,000	177,148	1969	325,270	1,524,109
1935	49,800	167,434	1970	193,361	956,891
1936	49,710	126,884	1971	441,332	1,841,519
1937	43,196	141,738	1972	521,037	1,814,277
1938	44,781	131,409		$8,492,798	$30,941,613
1939	47,171	136,202			
1940	43,827	146,726			
1941	50,815	192,177			
1942	49,856	138,953			
1943	60,364	137,641			
1944	105,885	203,433			
1945	78,006	187,909			

NOTE: A word of caution to anyone who may attempt to match federal funds with state appropriations appearing in another tabulation. Differences are the result of the year of actual receipt of federal funds. Work sheets are available.

3. EXPENDITURES - MAINE FORESTRY DISTRICT 1917

	St. John	Penobscot	Kennebec	Androscoggin	St. Croix Machias etc.	Total
Chief Wardens	$ 6,883.35	$ 4,946.11	$ 3,052.88	$ 315.32	$ 832.41	$16,030.07
Deputy Wardens	552.10	405.80	252.85	—	67.25	1,278.00
Lookout Expenses	4,219.26	6,934.80	10,394.71	2,290.28	3,749.33	27,588.38
Patrol Expenses	11,475.93	6,811.82	596.32	1,058.75	218.45	20,161.27
Fire Expenses	561.15	476.20	59.10	—	168.96	1,265.41
Tools and Supplies	1,683.28	2,836.50	2,782.78	225.49	1,208.70	8,736.75
Adm. Charges	426.09	426.01	426.95	440.48	436.29	2,155.82
Misc. Charges	79.80	93.02	79.80	79.79	79.81	412.22
	$25,880.96	$22,930.26	$17,645.39	$4,410.11	$6,761.20	$77,627.92

EXPENDITURES BY WATERSHEDS M.F.D. 1927

	St. John	Penobscot	Kennebec	Androscoggin	Machias	Total
Chief Wardens	$ 8,555.53	$ 6,864.75	$ 4,088.81	$ 743.98	$ 1,734.14	$ 21,987.21
Deputy Wardens		331.23		12.00	106.70	449.93
Lookout Stations	12,428.95	17,593.12	10,851.49	4,326.91	7,041.63	52,242.10
Patrolmen	12,299.11	7,437.47	2,210.68	2,605.35	2,799.68	27,352.29
Fire Fighting	1,942.95	1,401.91	25,833.41	2,924.61	928.33	33,031.21
Tools and Supplies	8,695.87	11,917.73	7,778.49	3,784.68	6,377.99	38,554.76
Administration	1,303.42	1,259.88	1,109.68	1,002.14	1,140.84	5,815.96
Miscellaneous	959.62	1,216.58	738.47	325.70	582.43	3,822.80
Totals	$46,185.45	$48,022.67	$52,611.03	$15,725.37	$20,711.74	$183,256.26

4. BUDGET AND OPERATING STATEMENT MAINE FORESTRY DISTRICT
January 1, 1950 to December 31, 1950

Acreage	Budget	Total Expenditures 10,262,455 A.	Augusta Office and Planes	Northern (Gilpatrick) 2,241,348 A.	Central (Pendleton) 2,532,467 A.	Eastern (Faulkner) 1,962,926 A.	Western (Hutton) 3,525,714 A.
Office Salaries	—	$ 11,348.08	$11,348.08	—	—	—	—
Supervisors	—	16,994.64	—	$ 4,123.60	$ 4,123.60	$ 4,623.84	$ 4,123.60
Chief Wardens	—	45,801.15	—	10,655.81	12,456.57	10,535.36	12,153.41
Watchmen	—	68,903.50	—	7,561.74	23,214.15	14,333.35	23,794.26
Patrolmen	—	81,197.69	—	22,941.09	17,404.45	14,901.46	25,950.69
Telephone Operators	—	4,167.00	—	1,278.00	1,328.00	376.00	1,185.00
Pilots	—	5,524.00	5,524.00	—	—	—	—
Total Personal Services	$238,000.00	233,936.06					
Plane Rentals	2,000.00	1,282.93	—	19.99	248.31	124.57	890.06
Plane Operation	—	3,560.04	3,560.04	—	—	—	—
Fire Suppression	49,000.00	49,877.78	—	3,652.42	8,849.85	30,718.74	6,656.77
Traveling Expenses	5,500.00	5,832.31	2,026.07	1,158.44	1,061.41	769.55	816.84
Car and Truck Operation	29,750.00	24,637.19	549.46	5,885.56	5,187.14	5,926.28	7,088.75
Utility Services	4,550.00	4,987.22	8.32	281.87	1,527.90	968.01	2,201.12
Rents	800.00	562.82	50.00	19.84	172.00	2.00	318.98
Repairs	10,000.00	12,221.34	11.37	2,816.88	2,578.73	2,227.64	4,586.72
Insurance	2,000.00	2,938.83	—	383.38	947.17	862.89	745.39
General Operating Expense	1,425.00	1,732.22	1,194.76	29.36	119.50	192.58	196.02
Food, Telephone and Repairs	700.00	1,114.58	—	—	458.41	19.90	636.27
Fuel	200.00	229.40	—	84.52	57.60	—	87.28
Office Supplies	950.00	323.38	271.26	—	25.43	16.86	9.83
Other Supplies	4,000.00	8,532.90	146.46	2,020.20	2,277.40	1,526.09	2,562.75
Disability Awards	2,548.00	1,574.18	1,134.00	20.50	33.00	14.00	372.68
Buildings and Improvements	7,750.00	7,514.55	—	3,132.48	2,432.68	483.66	1,465.73
Equipment	48,240.00	39,440.70	857.74	6,165.74	10,741.75	8,594.52	13,080.95
Radio Equipment	—	4,508.09	4,508.09	—	—	—	—
Equipment 1951	—	15,086.61	15,086.61	—	—	—	—
Exhibits	—	62.50	62.50	—	—	—	—
	$407,413.00	$653,891.69	$46,338.76	$72,231.42	$95,245.05	$97,217.30	$108,923.10

5. MAINE FORESTRY DISTRICT
EXPENDITURES
Fiscal Year Ending June 30, 1972

	Budget F.Y. 1972	Total Expenditures June 30, 1972	Augusta I. & E. Planes, etc.	NORTHERN DIVISION	EASTERN DIVISION	WESTERN DIVISION	Radio	Budget F.Y. 1973
Spruce Budworm Spray Proj.		319,898.26	319,898.26					
Fire Suppression Costs	60,000.	29,807.71	161.00	12,449.62	8,054.24	9,142.85		60,000.
Personal Services	860,995.	802,720.62	114,590.31	258,855.31	180,388.30	220,822.91	28,063.79	897,655.
Special Services	2,400.	3,477.42	2,233.37	239.30	549.26	455.49		2,410.
Travel Expenses	17,500.	21,346.77	7,728.99	5,499.57	3,288.34	4,412.89	416.98	17,500.
Operation of Vehicles	71,500.	74,324.34	20,739.57	28,641.36	19,659.49	23,562.20	1,721.72	72,500.
Operation of Planes	35,000.	35,651.31	35,651.31					35,000.
Utility Service	19,600.	20,713.68	1,017.11	6,019.74	7,105.93	6,397.82	173.08	19,600.
Rents	48,750.	50,644.97	1,683.69	14,543.60	10,251.01	25,166.67		49,750.
Repairs	33,950.	23,739.94	678.96	8,333.83	3,665.87	5,067.87	5,993.41	33,950.
Insurance	35,982.	31,313.57	19,412.47	3,377.71	3,652.39	3,779.65	1,091.35	35,982.
General Operating Expense	12,100.	12,141.81	9,543.07	844.20	678.31	1,072.23	4.00	12,100.
Food	500.	318.39	11.22	119.74	46.61	140.82		500.
Fuel	8,500.	6,286.35	1,249.51	2,113.27	1,898.05	1,025.52		8,500.
Office Supplies	4,050.	3,768.84	3,071.27	244.35	150.34	302.88		4,050.
Clothing	9,000.	10,016.04	9,880.66	74.00	17.76	43.62		9,000.
Supplies & Small Tools	22,550.	30,295.75	17,035.98	4,446.54	2,522.57	6,213.79	76.87	22,550.
Grants to Cities & Towns		2,520.77	2,520.77					16,000.
Purchase of land		2,500.00			2,500.00			
Buildings & Improvements	32,000.	6,470.64	644.82	345.37	41.63	5,438.82		82,000.
Equipment	138,510.	191,199.51	9,133.62	48,821.29	45,430.26	31,383.24	56,431.10	158,835.
Structures & Improvements	6,000.	3,224.34	1,774.69	6.65	1,443.00			9,000.
Retirement Contributions	101,225.	101,546.56	101,546.56					102,257.
TOTAL EXPENDITURES	1,520,112.	1,783,927.59	659,207.21	394,975.45	291,343.36	344,429.27	93,972.30	1,649,139.

6. REASONS FOR SIGNIFICANT VARIANCES
1971 BUDGET vs 1971 ACTUAL EXPENDITURES

Below follow examples of narrative comments in budget reviews for the M.F.D. Advisory Committee:

ITEM	VARIANCE (Actual to Budget)	REASON
Spruce Budworm Spray Proj.	$15,314.79 +	Radio equipment purchase for $10,010.00, ⅓ cost paid by state, ⅓ to be charged to federal government as rental on future projects. Spruce Budworm Film $6,600.00.
Fire Suppression Costs	15,371.35 +	Actual expenditures were less than budgeted in the Eastern Division and Headquarters, but were more for the Western and Northern Divisions.
Personal Services	60,740.02 −	The budget included maximum seasonal requirements whereas actual costs included $8,633.27 "netted" covering charges for Public Lot work, also there was a $9,020.00 savings on the Campsite Coordinator. The majority of the balance is savings through aircraft flights and personnel changes.
Operation of Vehicles	6,513.93 +	Increased use of vehicles for out-of-state use in obtaining excess property, increase costs of fuel, tires and general repair. The F.Y. 1972 budget has been adjusted to compensate for some of these costs.
Operation of Planes (1 Super Cub − 5 Helicopters − 6 Beavers & 2 Cessna 180)	5,261.13 +	This is only $2,859.40 over last years actual. This is due to increased uses, additional helicopters acquired late in the year and the actual budget increase was low. The budget for F.Y. 1972 has been increased.
Rents (Aircrafts)	5,270.41 +	New aerial detection flight added to Northern Division.
Insurance	8,145.80 −	This decrease is mostly in Workman's Compensation Insurance, whereas we had $20,000.00 budgeted and premiums were only $15,730.00 and a rebate of $2,977.00 (68–69 policy) was netted against this figure.

Clothing	7,430.25 −	Decrease due to carry-over of F.Y. 1970 inventory.
Supplies	14,814.76 +	$9,821.50 (of $19,643.00) was paid in F.Y. 1971 for filming of U.S. Forest Service movie "A Home for All Seasons" for which we were reimbursed in F.Y. 1972. Also paid for F.Y. 1970 obligations of fire prevention supplies from federal government.
Land	3,135.35 +	No funds are budgeted for land acquisition in that authority is given at the time of the required acquisition. This item covers the purchase of a forest rangers storehouse and site in the Town of Danforth.
Buildings & Improvements	1,634.80 +	Purchase of the former Gap Filler Annex on Musquash Mountain, Topsfield from the federal government for $3,500.00. By applying this against the increase, it shows that there was a decrease in this activity.
Equipment	16,626.86 +	This amount includes the purchase of 2 low-bed trailers, authorized but not budgeted and purchase orders carried from F.Y. 1970.
Structures & Improvements	3,616.22 +	This increase was due to construction of a concrete apron at the aircraft hangar in Old Town for helicopter landings and outside plane maintenance.

Appendix III

FOREWORD

The Northeastern Forest Fire Protection Commission is an unprecedented arrangement under which the affiliated member states, with the cooperation of the United States Forest Service, have joined in a united effort to prevent and control forest fires.

The objectives of the Commission are attained through the development of integrated forest fire plans, by the maintenance of adequate forest fire fighting services, by providing for mutual aid in fire suppression and by the establishment of a central executive office to plan and coordinate the services of the Commission.

This Reference Manual has been prepared as a cooperative undertaking by the Commission and the U. S. Forest Service, Region 7, as provided in Article VI of the Enabling Interstate Compact.

The purpose of this Manual is to provide a common reference for the methods and techniques of forest fire suppression in the northeast and a basis for training overhead personnel. It is designed to secure coordinated operations in interstate mutual aid fire suppression activities.

1958

Regional Forester, Region 7
U. S. Forest Service

Executive Secretary
Northeastern Forest Fire
Protection Commission

Appendix IV

1. FOREST FIRE RECORD - MAINE

(Figures rechecked, revised, and supersede all former tabulations)

Year	MAINE FORESTRY DISTRICT			ORGANIZED TOWNS			STATE TOTAL		
	No. Fires	Acreage	Damage	No. Fires	Acreage	Damage	No.	Acreage	Damage
1903	136	200,232	$ 761,588	209	67,355	$ 186,000	345	267,587	$ 947,588
1904	31	6,958	12,655		No Record		31	6,958	12,655
1905	109	14,737	40,518	33	5,579	23,105	142	20,316	63,623
1906	56	7,250	19,488	11	371	1,540	67	7,621	21,028
1907	16	2,324	5,257	17	2,200	9,310	33	4,524	14,567
1908	126	98,691	361,796	111	43,439	257,020	237	142,130	618,816
1909	68	27,083	63,734	89	11,945	32,965	157	39,028	96,699
1910	18	267	935	12	581	1,906	30	848	2,841
1911	127	99,654	289,052	75	11,423	48,303	202	111,077	337,355
1912	63	16,198	57,452	36	4,042	14,096	99	20,240	71,548
1913	74	9,327	28,477	120	20,887	148,365	194	30,214	176,842
1914	105	8,311	14,467	52	7,405	14,840	157	15,716	29,307
1915	80	14,472	22,776	76	11,185	55,340	156	25,657	78,116
1916	54	8,257	9,460	18	3,359	10,305	72	11,616	19,765
1917	19	147	1,334	9	311	800	28	458	2,134
1918	58	3,820	7,291	21	5,118	70,600	79	8,938	77,891
1919	85	4,352	6,305	19	668	2,625	104	5,020	8,930
1920	118	34,558	143,753	47	5,245	42,155	165	39,803	185,908
1921	250	56,947	404,555	112	11,883	112,560	362	68,830	517,115
1922	164	19,198	106,001	52	2,190	8,775	216	21,388	114,776
1923	132	62,407	289,845	49	7,932	51,521	181	70,339	341,366
1924	158	38,401	101,986	62	1,956	11,802	220	40,357	113,788
1925	73	2,328	14,058	42	3,725	29,060	115	6,053	43,118
1926	83	3,717	34,068	61	8,495	18,113	144	12,212	52,181
1927	60	9,096	103,649	49	2,524	25,705	109	11,620	129,354

FOREST FIRE RECORD - MAINE

(Figures rechecked, revised, and supersede all former tabulations)

Year	MAINE FORESTRY DISTRICT			ORGANIZED TOWNS			STATE TOTAL		
	No. Fires	Acreage	Damage	No. Fires	Acreage	Damage	No.	Acreage	Damage
1928	27	1,562	1,965	37	622	4,070	64	2,184	6,035
1929	90	1,323	11,363	78	1,142	33,394	168	2,465	44,757
1930	129	11,678	39,316	134	21,631	104,545	263	33,309	143,861
1931	92	562	1,580	134	4,245	51,417	226	4,807	52,997
1932	164	36,343	50,731	157	6,484	19,076	321	42,827	69,807
1933	165	5,299	7,259	116	9,995	41,568	281	15,294	48,827
1934	165	130,293	385,126	101	6,077	36,538	266	136,370	421,664
1935	220	14,582	28,001	81	4,246	9,557	301	18,828	37,558
1936	84	179	13,270	52	1,461	7,025	136	1,640	20,295
1937	162	1,358	12,191	100	4,355	18,023	262	5,713	30,214
1938	92	5,210	7,815	81	10,929	25,706	173	16,139	33,521
1939	128	2,914	15,757	159	4,519	20,953	287	7,433	36,710
1940	120	523	3,681	120	3,588	19,255	240	4,111	22,936
1941	157	12,847	82,543	324	27,503	428,797	481	40,350	511,340
1942	97	1,785	2,853	128	3,208	8,780	225	4,993	11,633
1943	37	244	4,157	94	6,924	35,753	131	7,168	39,910
1944	147	12,162	121,773	261	12,041	157,091	408	24,203	278,864
1945	83	889	4,590	131	4,061	59,993	214	4,950	64,583
1946	151	3,553	29,482	425	6,774	66,450	576	10,327	95,932
1947	167	4,685	20,167	533	208,862	11,970,688	700	213,547	11,990,855
1948	266	805	33,075	548	6,436	102,358	814	7,241	135,433
1949	219	16,938	27,618	544	4,114	38,223	763	21,052	65,841
1950	172	6,516	71,334	779	11,535	92,457	951	18,051	164,791
1951	118	503	10,492	303	3,182	20,489	421	3,685	30,981
1952	301	18,615	535,899	647	6,080	90,052	948	24,695	625,951

258

Year									
1953	200	4,886	29,506	677	9,672	90,549	877	14,558	120,055
1954	75	453	3,676	266	2,727	18,846	341	3,180	22,522
1955	168	528	16,362	322	1,254	13,115	490	1,782	29,477
1956	128	552	2,392	315	2,028	10,684	443	2,580	13,076
1957	174	5,745	81,128	552	25,222	235,149	726	30,967	316,277
1958	65	364	3,448	156	1,198	8,228	221	1,562	11,676
1959	137	1,958	26,989	348	4,613	30,245	485	6,571	57,234
1960	160	1,802	112,357	312	1,008	15,945	472	2,810	128,302
1961	132	926	17,094	270	1,555	10,171	402	2,481	27,265
1962	111	1,079	31,241	352	2,359	68,302	463	3,438	99,543
1963	207	1,224	39,266	337	584	18,734	544	1,808	58,000
1964	211	1,237	10,685	493	2,345	9,918	695	3,582	20,603
1965	212	13,008	*	761	3,472	—	972	16,480	751,000
1966	236	331	—	344	3,030	—	580	1,360	—
1967	106	980	—	241	639	—	347	1,619	—
1968	211	4,739	—	305	1,509	—	516	6,248	—
1969	86	308.7	—	214	2,089.7	—	300	2,398	—
1970	158	494	—	272	517	—	430	1,011	—
1971	230	247	—	246	520	—	476	767	—
1972	103	322	—	327	1,330	—	430	1,652	—
TOTAL	8,916	1,081,283.7	$4,796,682	14,559	685,503.7	$15,169,955	23,475	1,766,786 A	$20,717,637
	37.9%	61.2%	23%	62.1%	38.8%	67%			

*No damage figures since 1965 as the U.S. Forest Service has been working on a better method of determining more accurate figures in this category. Data will be worked back to fill in missing years.

2. Notes from Diary of Elmer B. Crowley, Civil Engineer

Wednesday, May 24, 1905 –
I made a pencil plan of the lot at Moosehead and one of W. M. Shaw's house lot. Begun on a plan for the use of a fire warden on *Squaw Mountain*, at 3:00 p.m.

Thursday,
Friday,
Saturday and Monday –
Worked on this plan and printed it.

Monday and Tuesday,
Made a rig to take up on the mountain.

Wednesday –
Went to Camp No. 2 and got dinner. I had a black mare and took all my things up. In P.M. I started for the summit with drills, hammer, axe, big board and about 3 feet of 2" pipe, and miscellaneous things. I got all but the board to the top of the mountain. I drilled 4 holes about 4" into the ledge at about 20 feet south of the U. S. bench mark. I then drove in pine plugs and with 3" lag bolts fastened a collar to the ledge. I then threaded in the pipe and built a leanto of tarred paper over the site and went back to *Camp No. 2*.

Thursday –
I went to the top of the mountain with the board map. I set up and sighted at the different points to check up on my map. It checked very well excepting Moose Island and Greenville Village and Squaw Ponds. It was very hazy and, of course, I had to guess at the distances. I got dinner at No. 2 and came to Greenville accompanied by John M. Conley. I reached there about 3:45 p.m. I left the board under the leanto on the mountain, made out a bill for $31.25 of which $1.50 was paid out for myself.

Saturday, June 10 –
Went and got *Will Hilton* established on Squaw Mountain and returned to Greenville at 7:00 p.m.

Wednesday, June 14 –
At first I made 2 prints of the fire plan and got one ready to send to *Mr. Ring*.

Wed., and Thurs., August 30, and 31, 1905
On Wednesday I took the turn at Long Pond and went to Attean Landing. Mr. Newton joining me at Jackman. We went on to Attean Mountain from which I took sights at mountains and ponds etc., for the purpose of making a fire plan to be used on Attean Mountain. I stayed at the Newton House and came to Greenville on Thursday.

On Friday and Saturday I worked on the above plan with the exception of the time for making 2 prints of the Kennebec Log Driving Company's lot at Moosehead.

Observations taken from Attean Mountain
Highest mountain on Holeb line N.44W.
Owls Head S.74E.
Mountain on 3R4
or R7 S.57-30E.
Williams' Mountain S.51-30E
Burnt Jacket N.27W.
East end Little Big Wood Pond N.10-15E.
North end Big Wood Pond N.10-50E.
Sandy Stream slightly off N6-30W
Sandy Stream Mountain N12E
West end Bald Mountain N22-40E

Monday, Sept. 4, 1905

I worked on the Attean fire plan until 12:00 noon. My time was broken into somewhat.

Tuesday, I worked on the Armstrong Lot until 2:30 p.m., then worked on the Attean fire plan until 6:00 p.m.

Monday night, May 5, 1906

I went to Jackman

Tues., I went to Attean Mountain and took notes of reference points to be used on the fire plan.

Thursday, May 24, 1906

I worked 2 hours in the wiring on the Attean fire plan.

Tues., May 29, 1906

In the evening I traced 2 hours on the Attean plan.

Wed., May 30, 1906

Wednesday evening I worked 2½ hours and in the day, Friday, 4 hours and 2 hours on Saturday.

April 23, 1906 Monday and Tuesday

Made a pencil plan for the Whitecap Mountain Fire Station.

April 29, 1906 Sunday

Went to Whitecap Mountain and took observations on Sunday. The wind blew so hard between 10:00 and 11:00 A.M. it was a hard chance. On the north slope, I found six feet of snow and the top of the mountain had snow to the tops of the stunted spruces. There will have to be two stands for observation points unless a tower is built. The data on pages 17 and 18 refer to the numbers of the pencil plan. I set up my plan in approximately the right position and sighted these places and marked them off on the map. Those on page 17 were taken on the east side and those on page 18 were on the west side.

Page 17:

1.	Head of Third Roach Pond	19.	South Shore of Ponds
3.	West point of Big Spencer Mountain	22.	Shore of Pond
4.	East point of Big Spencer Mountain	26.	Center of Silver Lake
5.	North inlet	28.	Big Spruce Mountain
6.	Shaw Mountain	29.	Big Houston Pond
7.	Pond	30.	Sebec Lake
8.	Ragged Lake	31.	Barren Mountain
9.	Ragged Lake	32.	Lake "Link of Ponds"
14.	Katahdin	34.	End of Moose Island
16.	Boardman Mountain	35.	High part of Baker Mountain
17.	North end of Joe Mary Pond	36.	Rum Mountain
18.	High Mountain	37.	Elephant Head

May 2 & 3, 1906   Wednesday and Thursday
Worked on Fire Plan of Whitecap Mountain.

May 18, 1906   Friday
Traced on Whitecap Mountain Fire Plans in P.M. and two hours on Saturday.

June 3 & 4, 1906   Sunday and Monday
Worked four hours on Whitecap Mountain Fire Plan.

3.   As of 1973 the following are the remaining thirteen key operational towers in the M.F.D., maintained to provide air to ground communications. Possibly more will be eliminated as time goes on.

Northern Region
1. Allagash Mt., T7 R14, WELS, Piscataquis County
2. Rocky Mt., T17 R12, WELS, Aroostook County
3. Ross Mt., T11 R15, WELS, Aroostook County
4. DeBoulie, T15 R9, WELS, Aroostook County
5. Number 9 Mt., T.D. R2, WELS, Aroostook County
6. Round Mt., T11 R8, WELS, Aroostook County

Eastern Region
1. Cooper Mt., Cooper, Washington County
2. Almanac Mt., Lakeville Plt., Penobscot County
3. Ragged Mt., Indian #4, Penobscot County

Western Region
1. West Kennebago, T4 R4, WEKP, Oxford County
2. Mt. Bigelow, Dead River Plt., Somerset County
3. Spencer Mt., T2 R13, WELS, Piscataquis County
4. Green Mt., T4 R13, WELS, Somerset County

   As stated earlier in the building and expanding process of perfecting an efficient network of lookout towers, many of them were either relocated or abandoned. Scattered throughout the records are accounts of some of them. For the useful purpose they once served, it seems worthwhile to recognize most of them again collectively. (Some may have been overlooked or may be missing from the records.)

Some Older Towers Abandoned Many Years Ago

Attean Mt., Attean Twp., Somerset County
Sally Mt., T5 R1, Somerset County
Flagstaff Mt., Flagstaff Plt., Somerset County
Mucakea Mt., T5 R16, Somerset County
Picket Mt., Lang Plt., Somerset County
Cobb Hill, Lee, Penobscot County
Horse Mt., T6 R8, Penobscot County[1]
Double Top, T3 R10, Piscataquis County[1]
Center Mt., T4 R10, Piscataquis County[1]
Pogey Mt., T4 R9, Piscataquis County
Mt. Katahdin, T3 R9, Piscataquis County[1,2]
Joe Mary Mt., TA R10, Piscataquis County
Black Cat, T1 R9, Piscataquis County
Lily Bay Mt., TA R14, Piscataquis County
Burnt Mt., T5 R10, Piscataquis County
City Camps, T4 R9, Piscataquis County
Boarstone Mt., Elliottsville Plt., Piscataquis County
Kineo Mt., Days Academy Grant, Piscataquis County
Tug Mt., T30 M.D., Washington County

---

[1] All are within the present limits of Baxter State Park but originally outside on private land.

[2] Mt. Katahdin lookout was never a tower but only a cabinlike structure. Erected and maintained 1913-1919 and abandoned in 1920 due to too much fog, clouds and haze. Philip T. Coolidge, prominent Maine consulting forester, now deceased, was first visitor to register in 1914 at watchmans camp located on Abol trail below slide.

MAINE FORESTRY DISTRICT
Lookout Stations Operated in 1917

NO.	NAME	LOCATION	CHIEF WARDEN	WATCHMEN
* 1	Lead Mountain	Twp. 28 Hancock Co.	F.E. Patten, Cherryfield	Hiram Corliss, Cherryfield
2	Pleasant Pond Mt.	Caratunk Plantation	John B. Comber, Caratunk	Willie Williams, Caratunk
* 3	Attean Mountain	Attean Twp.	G.G. Nicholas, Jackman	Allan Runnels, Jackman
* 4	Tumbledown Mt.	Twp. 5 R. 6 W.K.R.	Ralph Wing, Flagstaff	Fred L. Hutchins, Stratton
5	Squaw Mt.	Twp. 2, R. 6 E.K.R.	Louis Oakes, Greenville Jct.	Frank P. Conley, Greenville Jct.
6	Snow Mt.	Twp. 2, R. 5, Franklin Co.	Ralph Wing, Flagstaff	Leon Foster, Eustis
7	Mt. Bigelow	Bigelow Twp. Somerset	Ralph Wing, Flagstaff	L. F. Marsh, Dead River
8	White Cap Mt.	7 R. 10 N.W.P.	J.L. Chapman, Milo	A. J. Smart, Milo
9	Spencer Mt.	Middlesex Grant, Piscataquis	R.L. Brick, Levant	E. S. Turner, S. D. Call, Levant
10	Rocky Mt.	Twp. 18, R. 12, W.E.L.S.	Harry E. Hasey, St. Francis	Fred A. Lancaster, Old Town
11	Pogey Mt.	Twp. 4, R. 8, W.E.L.S.	John E. Mitchell, Patten	Andrew Finnegan, Patten
12	Otter Lake Mt.	Twp. 3, R. 4, W.E.L.S.	H.G. Tingley, Island Falls	Gilbert Scoville, Island Falls
*13	Mt. Chase	Chase Twp.	John E. Mitchell, Patten	Warren Darling, Smyrna Mills
*14	Ragged Mt.	Twp. A. R. 9, W.E.L.S.	E.M. Chase, Brownville	George Monroe, Milo
*15	Mt. Kineo	Moosehead Lake	Louis Oakes, Greenville Jct.	E. J. Conley, Greenville Jct.
16	Mt. Coburn	Twp. 3, R.6, B.K.P. W.K.R.	John B. Comber, Caratunk	Melville Blethen, Foxcroft
17	Wesley Mt.	Wesley, Washington Co.	Herbert M. Gardner, Machias	C. M. Archer, Wesley
18	Depot Mt.	Twp. 14, R. 16, W.E.L.S.	Grover C. Bradford, Sebec	Clyde Fox, St. Pamphile, P.Q.
19	Soper Mt.	Twp. 8, R. 12, W.E.L.S.	Eugene H. Decker, Old Town	F.L. Berry, Bangor
20	Round Mt.	Twp. 11, R. 8, W.E.L.S.	Chas. L. Weeks, Masardis	Ira D. McKay, Ashland
21	Aziscoos Mt.	Lincoln Plantation, Oxford	S. F. Peaslee, Upton	Calvin T. Fox, Wilson's Mills
22	Mt. Katahdin	Twp. 3, R. 9, W.E.L.S.	Thos. Griffin, Millinocket	Frank Sewall, Millinocket
*23	Bald Mt.	Twp. 3, R. 3, E.K.R.	Frank Hilton, Bingham	Richard Morris, The Forks
24	Kibbie Mt.	Twp. 1, R. 6, W.E.L.S.	L. P. Barney, Skinner	Louis LeRoy, Tarratine
25	Priestly Mt.	Twp. 10, R. 13, W.E.L.S.	Harry E. Hasey, St. Francis	Roy Stewart, St. Francis
26	Boundary Bald	Twp. 4, R. 3, W.E.L.S.	George Nichols, Jackman	Philander McKenney, Jackman
*27	Williams Mt.	Twp. 2, R. 7, B.K.P., W.K.R.	L. P. Barney, Skinner	Herbert Holden, Tarratine
*28	West Kennebago	Twp. 4, R. 14, W.E.L.S.	C. C. Murphy, Rangeley	
29	No. 4 Mt.		Louis Oakes, Greenville Jct.	Mahlon G. Coughlin, Kokadjo
30	Cobb Hill	Lee, Penobscot Co.	Leroy Brown, Lee	Earl Ware, Lee
31	Taylor Hill	Princeton, Washington	Geo. E. Andrews, Princeton	James S. Kneeland, Princeton
*32	Tug Mt.	Twp. 30 M.D. Washington	Herbert M. Gardner, Machias	John Roberts, Machias
33	Beetle Mt.	Twp. 7, R. 10, W.E.L.S.	John E. Mitchell, Patten	William Finch, Patten
34	Mattagamon Sta.	Twp. 6, R. 8, W.E.L.S.	John E. Mitchell, Patten	Joseph Mitchell, Patten
*35	Boarstone Mt.	Elliotsville Plantation	J. L. Chapman, Milo	F. H. Small, Onawa
*36	Joe Mary Mt.	Twp. A. R. 10	Thos. Griffin, Millinocket	C. B. Wood, Waterville
*37	Cooper Mt.	Cooper, Washington	Geo. E. Hathaway, Jacksonville	Oscar Sadler, Cooper
*38	Musquash Mt.	Topsfield	Geo. E. Andrews, Princeton	Warren A. Bailey, Waite
39	Green Mt.	Twp. 4, R. 18, W.E.L.S.	D. H. Lambert, Old Town	Earl Ford, Seboomook
40	Mucalsea Mt.	Twp. D, R. 16, W.E.L.S.	D. H. Lambert, Old Town	John Ford, Seboomook
41	Saddleback Mt.	Twp. 5, R. 1, W.B.K.P.	C. C. Murphy, Rangeley	Kenneth Lee, Augusta
42	Double Top Mt.	Twp. 4, R. 10, W.E.L.S.	Thos. Griffin, Millinocket	F. L. Sawyer, Millinocket
43	Nulhedus Mt.	Twp. 5, R. 17, W.E.L.S.	D. H. Lambert, Old Town	Thomas Fleming, Bangor
44	Lawler Hill	Benedicta	Rex E. Gilpatrick, Davidson	
45	Norway Bluff	Twp. 9, R. 9, W.E.L.S.	Chas. L. Weeks, Masardis	Fred Johnston, Masardis
46	No. 9 Mt.	Twp. D, R. 2, W.E.L.S.	James M. Pierce, Houlton	W. B. Hussey, Patten
*47	Hedgehog Mt.	Twp. 14, R. 6, W.E.L.S.	John M. Brown, Eagle Lake	Herbert E. Brown, Eagle Lake
48	Three Brooks Mt.	Twp. 15, R. 6, W.E.L.S.	Claude M. Austin, Old Town	John M. Donahue, Guerette
49	Speckles Mt.	Grafton, Oxford	S. F. Peaslee, Upton	Harry Noyes, Bryant's Pond
50	Spoon Mt.	Twp. 8, R. 7	John E. Mitchell, Patten	Joseph Ingraham, Patten
*51	Mattamiscontis Mt.	Twp. 3, R. 9, N.W.P.	E. M. Chase, Brownville	John Stinchfield, Milo
*52	Haystack Mt.	Twp. 11, R. 4, W.E.L.S.	Chas. L. Weeks, Masardis	M. H. Friedman, Presque Isle
*53	Schoodic Mt.	Twp. 9, Hancock Co.	Fred S. Bunker, Franklin	Howard Webb, No. Sullivan
54	Hardwood Mt.	Twp. 9, R. 18, W.E.L.S.	Wm. H. Hinckley, St. Camille P.Q.	Edwin Costello, Bangor
55	Almanac Mt.	Lakeville Plantation	Leroy Brown, Lee	
56	Allagash Mt.	Twp. 7, R. 14, W.E.L.S.	Eugene Decker, Old Town	John Furey, Bangor

*Federal Service - 22. The Federal allotment, as in previous years, was confined entirely to Lookout Stations, selected by the Chief of State Co-operation.

5. LOOKOUT STATIONS OPERATED IN 1932

ST. JOHN WATERS
1 Three Brooks Mt. in Twp. 15, R. 6
2 Stockholm Mt. in Stockholm
3 Carr Pond Mt. in Twp. 13, R. 8
4 Hedgehog Mt. in Twp. 15, R. 6
5 DeBoulie Mt. in Twp. 15, R. 9
6 Rocky Mt. in Twp. 17, R. 12
7 Musquacook Mt. in Twp. 14, R. 12
8 Depot Mt. in Twp. 14, R. 16
9 Hardwood Mt. in Twp. 9, R. 18
10 Squapan Mt. in Twp. 11, R. 4
11 Norway Bluff in Twp. 9, R. 9
12 Round Mt. in Twp. 11, R. 8
13 Oak Hill in Twp. 8, R. 5
14 Howe Brook Mt. in Twp. 8, R. 3
15 No. 9 Mt. in Twp. D, R. 2
16 Clear Lake Mt. in Twp. 10, R. 11
17 Priestly Mt. in Twp. 10, R. 13

KENNEBEC WATERS
18 Mt. Kineo in Day's Academy
19 No. 4 Mt. in Twp. A, R. 13
20 Squaw Mt. in Twp. 2, R. 6
21 Wadleigh Mt. in Twp. 1, R. 12
22 Bigelow Mt. in Dead River Pl.
23 Snow Mt. in Twp. 2, R. 5
24 Flagstaff Mt. in Flagstaff Pl.
25 Mt. Abram in Twp. 4, R. 1
26 Mt. Coburn in Twp. 3, R. 6
27 Moxie Bald Mt. in Twp. 2, R. 3
28 Pleasant Pond Mt. in The Forks Pl.
29 Boundary Bald Mt. in Twp. 4, R. 3
30 Sally Mt. in Twp. 5, R. 1
31 Williams Mt. in Twp. 2, R. 7
32 Kibbie Mt. in Twp. 1, R. 7
33 Tumbledown Mt. in Twp. 5, R. 6

ANDROSCOGGIN WATERS
34 West Kennebago Mt. in Twp. 4, R. 4
35 Aziscoos Mt. in Lincoln Pl.
36 Old Spec Mt. in Grafton

37 Saddleback Mt. in Sandy River Pl.
38 Deer Mt. in Twp. 4, R. 2

PENOBSCOT WATERS
39 Green Mt. in Twp. 4, R. 18
40 Little Russell Mt. in Twp. 5, R. 16
41 Nulhedus Mt. in Twp. 4, R. 17
42 White Cap Mt. in Twp. 7, R. 10
43 Mattamiscontis Mt. in Twp. 3, R. 9
44 Boarstone Mt. in Elliottsville Pl.
45 Trout Mt. in Twp. 2, R. 9
46 Ragged Mt. in Indian No. 4
47 Horse Mt. in Twp. 6, R. 8
48 Beetle Mt. in Twp. 7, R. 10
49 Mt. Chase in Twp. 8, Mt. Chase
50 Spoon Mt. in Twp. 8, R. 7
51 Burnt Mt. in Twp. 5, R. 10
52 Spencer Mt. in Twp. 2, R. 13
53 Soubunge Mt. in Twp. 4, R. 11
54 Doubletop Mt. in Twp. 3, R. 10
55 Soper Mt. in Twp. 8, R. 12
56 Allagash Mt. in Twp. 7, R. 14
57 Passadumkeag Mt. in Grand Falls Pl.
58 Dill Ridge in Lakeville Pl.
59 Mitchell Mt. in Haynesville
60 Otter Lake Mt. in Twp. 3, R. 4
61 Whitney Hill in Macwahoc Pl.
62 Daicey Mt. in Twp. 3, R. 7
63 Lawler Hill in Twp. 2, R. 6

ST. CROIX, MACHIAS, NARRAGUAGUS AND UNION WATERS
64 Schoodic Mt. in Twp. 9, S.D.
65 Musquash Mt. in Topsfield
66 Pirate Hill in Twp. 11, R. 3
67 Pocomoonshine Mt. in Princeton
68 Cooper Mt. in Cooper
69 Lead Mt. in Twp. 28, M.D.
70 Wesley Mt. in Wesley
71 Washington Bald Mt. in Twp. 42, M.D.
72 Peaked Mt. in Twp. 30, M.D.

6. LOOKOUT TOWERS - MAINE FORESTRY DISTRICT

Name of Tower	Location	County	Year tower first established	Material	Tower height (feet)	Year of Replacement	Material	Tower height (feet)	Elevation of Mt. (feet)
Abram	Mt. Abram, BKP WKR	Franklin	1924 (Rebuilt in 1926)	Steel	20				4,049
Allagash	T 7, R 14 WELS	Piscataquis	1916	Wood		1924	Steel	27	Unsurveyed
Aziscoos	Lincoln Pl.	Oxford	1910	Wood		1929	Steel	24	3,215
Beetle	T 7, R 10, WELS	Piscataquis	1913 (Rebuilt in 1917 - wood) (Relocated in 1919 - wood)	Wood	12				Unsurveyed
Bigelow	Dead River Pl.	Somerset	1905	Wood		1917	Steel	38	4,088
Borestone	Elliottsville Pl.	Piscataquis	1913 (House on ledge)	Wood					1,600
Boundary Bald	T 4, R 3, NBKP	Somerset	1911 (Rebuilt in 1914 - wood)	Wood		1937	Steel	35	3,000
Burnt	T 5, R 10, WELS	Piscataquis	1924	Steel	40		(Steel from Naval Radio Tower, Bar Harbor)		Unsurveyed
Carr Pond	T 13, R 8, WELS	Aroostook	1925	Steel	48				1,390
Chase	Mt. Chase Twp.	Penobscot	1909	Wood		1917	Steel	16	Unsurveyed
Clear Lake	T 10, R 11, WELS	Piscataquis	1929	Steel	24	1914	Steel	24	1,855
Coburn	T 3, R 6, BKP WKR (Upper Enchanted)	Somerset	1910	Wood			Steel (Cab rebuilt in 1938 - crushed by ice)		3,718
Cooper	Cooper	Washington	1913 (Rebuilt in 1925 - wood)	Wood		1937	Steel (Steel from Naval Radio Tower, Bar Harbor)	80	Unsurveyed
Deasey	T 3, R 7, WELS	Penobscot	1929 (House on ledge)	Wood					Unsurveyed
DeBoullie	T 15, R 9, WELS	Aroostook	1920-21	Steel	12	1929-30	Steel (Cab rebuilt in 1937 - struck by lightning)	48	1,898
Deer	T 4, R 2, WBKP (Crockertown)	Oxford	1926	Steel	39				3,455
Depot	T 14, R 16, WELS	Aroostook	1909	Wood		1914	Steel	60	1,300
Dill Ridge	Lakeville Pl.	Penobscot	1927	Steel	48				948
Doubletop	T 3, R 10, WELS	Piscataquis	1913	Wood		1917-18	Steel	48	3,600
Flagstaff	Flagstaff Pl.	Somerset	1917	Steel	50				2,497
Green	T 4, R 18, WELS	Somerset	1913	Logs		1920	Steel (Relocated from east to west peak)	48	1,500
Hardwood	T 9, R 18, WELS	Somerset	1916	Steel	75				1,300
Hedgehog	T 15, R 6, WELS	Aroostook	1914	Steel	24				1,594
Horse	T 6, R 8, WELS	Penobscot	1917	Steel	15				Unsurveyed
Horseshoe	T 11, R 10, WELS	Aroostook	1935	Wood	20				2,052
Howe Brook	T 8, R 3, WELS	Aroostook	1930	Steel	75				1,458
Indian Hill	Grand Lake Stream Pl.	Washington	1934	Wood	30				782
Kibbie	T 1, R 7, WBKP (Skinner Town)	Franklin	1906 (Rebuilt in 1914 - steel)	Wood		1926	Steel	14	3,638
Kineo	Days Academy Grant	Piscataquis	1910 (Lookout house only)	Wood		1917-18	Steel	64	1,800
Lawler Hill	T 2, R 6, WELS	Penobscot	1914	USGS Poles	28	1931	Steel (Relocated from Hunt Mt.)	60	
Lead	T 28 MD	Hancock	1910	Wood		1914	Steel	36	1,475
Little Russell	T 5, R 16, WELS	Somerset	1920	Steel	48				Unsurveyed
Mattamiscontis	T 3, R 9, NWP	Penobscot	1914	Wood		1917	Steel	48	1,400
May	Island Falls	Aroostook	1920	Steel	48				920
Millinocket Hill	Millinocket	Penobscot	(Erected by landowners - purchased by MFD - 1942) 1934	Wood	30				Unknown

Name	Location	Township	Notes	Built	Material	Height	Elevation
Mitchell	Haynesville	Aroostook		1918	Wood	36	567
Moxie Bald	T 2, R 3, BKP EKR	Somerset		1910	Wood		2,630
Musquacook	T 14, R 12, WELS	Aroostook		1925	Steel	60	1,500
Musquash	Topsfield	Washington		1913	Wood		Unsurveyed
Norway Bluff	T 9, R 9, WELS	Piscataquis		1914	Steel	24	Unsurveyed
Nulhedus	T 4, R 17, WELS	Somerset		1914	Steel	60	Unsurveyed
Number 4	T A, R 13, WELS (French Town)	Piscataquis		1913	Wood		
Number 5	T 6, R 7, BKP WKR (Appleton)	Somerset		1933	Steel	47	3,168
Number 9	T D, R 2, WELS	Aroostook	(New tower built in 1915 - wood)	1914	Wood		1,638
Oak Hill	T 8, R 5, WELS	Aroostook		1924	Steel	75	1,096
Old Spec	Grafton	Oxford		1914	Wood		4,250
Otter Lake	T 3, R 4, WELS	Aroostook	(Only lookout trees)	1911–12	Steel		595
Passadumkeag	Grand Falls Pl.	Penobscot		1919	Steel	36	1,463
Peaked	T 30, MD	Washington		1931–32	Steel	36	1,200
Pirate Hill	T 11, R 3, NBPP	Washington		1925	Steel	60	
Pleasant Pond	The Forks Pl.	Somerset		1910	Wood	58	2,480
Pocomoonshine	Princeton	Washington		1917	Wood	73	
Priestly	T 10, R 13, WELS	Piscataquis	(Caboose on top of 12 logs)	1910	Wood	22	1,900
Ragged	Indian No. 4	Penobscot	(Built by landowners)	1909–10	Wood	36	1,303
Rocky	T 17, R 12, WELS	Aroostook	(Rebuilt in 1917 – wood)	1907	Wood	48	1,400
Round	T 11, R 8, WELS	Aroostook	(Rebuilt in 1916 – wood)	1909	Wood	48	2,147
Sabao	T 41, MD	Hancock		1937	Steel	36	1,087
Saddleback	Sandy River Pl.	Franklin	(Cab rebuilt in 1938 – crushed by ice)	1913	Steel	36	4,116
Schoodic	T 9, SD	Hancock		1914	Steel	24	1,069
Snow	T 2, R 5, WBKP	Franklin		1910	Steel	24	3,948
Soper	T 8, R 12, WELS	Piscataquis		1909	Steel	27	Unsurveyed
Soubunge	T 4, R 11, WELS	Piscataquis	(Rebuilt in 1916 – wood)	1918	Steel	12	
Spencer	T 3, R 13, WELS	Piscataquis		1906	Steel	12	Unsurveyed
Spoon	T 8, R 7, WELS	Penobscot		1916	Steel	50	3,035
Squa Pan	T 11, R 4, WELS	Aroostook	(15' added in 1920)	1911–18	Wood		Unsurveyed
Squaw	T 2, R 6, BKP EKR	Piscataquis	(Log cabin structure)	1905	Wood	30	
				1926	Steel	48	1,460
				1919	Steel	12	3,209
Stockholm	Stockholm WELS	Aroostook		1924	Steel	75	974
Three Brooks	T 15, R 6, WELS	Aroostook		1914	Steel	48	1,578
Trout	T 2, R 9, WELS	Piscataquis	(Relocated from Black Cat Mt)	1931	Steel	60	1,420
Tumbledown	T 5, R 6, BKP WKR	Somerset		1910	Wood		
Wadleigh	T 11, R 12, WELS	Piscataquis		1927	Steel	36	3,542
Washington Bald	T 42, MD	Washington		1918	Wood	55	1,000
Wesley	Wesley	Washington		1910	Wood		1,100
			(Cab rebuilt in 1937 – struck by lightning)				
			(Steel from Naval Radio Tower, Bar Harbor)				Unsurveyed
West Kennebago	T 4, R 4, WBKP	Oxford		1911	Wood	24	3,705
White Cap	T 7, R 10, NWP (Bowdoin College Grant)	Piscataquis	(Rebuilt in 1914 – peeled logs 20')	1906	Steel	24	3,707
Whitney Hill	Macwahoc Pl.	Aroostook		1929	Steel		610
Williams	T 2, R 7, BKP WKR (Misery)	Somerset		1911	Wood	48	2,395

7. COOPERATIVE LOOKOUT TOWERS* - 1952

Name of Tower	Tel. Ex.	Location	Ownership	Name of Watchman	P.O. Address
Green Mt.	Center Ossipee, N.H., 8121-13	Effingham, N.H.	Maine	Harold Linscott	RFD, Center Ossipee, N.H.
Speckled Mt.	Gorham, N.H. 49-13	Stoneham, Me.	U.S.Forest Service	Don Starbird	Send to U.S.F.S. Dist. Hdqts.- So.Paris, Me. Norway-Paris 648-M
Pequawket Mt.	No. Conway, N.H. 8407-3	Conway, N.H.	U.S. Forest Service	Winn Whitman	Sent to U.S.F.S. Rangers' Office Conway, N.H. 109-2
Bald Mt.	Calais 672	Baring Twp., Me.	U.S. Wildlife Refuge	James W. Gillespie	U.S. Wildlife Refuge, Calais, Maine
Thompson Mt.	Call St. Eleuthere Central Office	Chabot, P.Q.	St. Lawrence Protective Assoc.	Isidore St. Pierre	
Beech Mt.	Southwest Harbor 310	Southwest Harbor Maine	Acadia Nat'l Park	Richard Young	Bar Harbor
Blue Job Mt.	Rochester, N.H. 852-5	Farmington, N.H.	New Hampshire	Harold E. Flower	RFD, Rochester, N.H.
Milan Hill	Milan, N.H. 2061	Milan, N.H.	New Hampshire	Delon Niclason	Milan, N.H.
Magalloway Mt.	Pittsburg, N.H. 8017-2	Pittsburg, N.H.	New Hampshire	Charles Heath	Pittsburg, N.H.
Deer Mt.	Pittsburg, N.H. 19-6	Pittsburg, N.H.	New Hampshire	Harlan F. Gilkey	Pittsburg, N.H.
Ste. Lucie		Talon, P.Q.	Canadian F.P. Serv.	Alfred Turgeon	
Littles Mt.	Pembroke 32-12	Edmunds Twp.	U.S. Wildlife Refuge	Dale Smith	Dennysville RFD No. 1

*There were other cooperative agreements from time to time as lookout towers were abandoned or relocated

Appendix V

1.

EQUIPMENT INVENTORY SUMMARY - 1972

	Northern Region	Eastern Region	Southern Region	Western Region	State Totals 1970	State Totals 1972
1" Linen Hose (feet)	51,100	45,000	37,700	41,600	171,050	175,400
1" S.J.R.L. Hose (feet)	5,200	2,900	500	2,200	7,240	10,800
1½" Linen Hose (feet)	162,800	148,800	82,200	134,900	521,600	528,700
1½" R.L. Hose (feet)	4,000	16,300	29,800	10,100	61,600	60,200
Power Pumps	110	96	60	98	350	364
Trailer Pumps	5	10	2	4	26	21
Indian Pumps - Hand	836	910	765	801	3,584	3,312
Axes	889	770	537	909	3,327	3,105
Shovels	904	674	475	730	3,006	2,783
Miscellaneous Hand tools	480	743	698	463	2,590	2,384
Boats	24	12	2	19	61	57
Canoes	33	17	-	13	64	63
Outboard Motors	48	32	2	31	107	113
Radios - mobiles	63	45	38	48	154	194
Radios - Battery Powered	82	47	63	60	298	252
Radios - AC Powered	15	18	20	15	70	68
Hose Packs	452	389	252	434	1,596	1,527
Basket Skidder	7	5	-	8	-	20
Tank Skidder	9	3	-	7	-	19
Relay Tanks	35	26	4	31	116	96
Portable Tanks (Harodikes)	48	26	11	21	96	106
Chain Saws	41	32	15	25	110	113
House Trailers	7	3	2	4	21	16
Trucks or Jeeps	88	94	75	81	355	338
Cars	4	2	4	2	12	12
Bulldozers	2	4	3	2	17	11
Fire Plows	-	4	3	-	10	7
Planes	4	2	1	2	8	9
Helicopters	1	2	1	1	2	5

Note: Miscellaneous Hand Tools include: Pick Axes, Bush Hooks, Hazel Hoes, Mattocks, Fire Rakes, and McLeod Tool.
Axes include: Pulaski, Forestry, Single bit, and Double bit.
Shovels include: D.H.R.P., L.H.R.P., and Forestry Ladies.

2.

CAPITAL VALUATIONS
REAL ESTATE AND CONTENT INFORMATION
(1972)

REAL ESTATE	BUILDING VALUE	CONTENT VALUE
Forest Nursery	92,500	86,000
Organized Towns	751,000	432,900
Forest Management	75,800	32,200
Entomology	50,000	55,600
Maine Forestry District	908,300	876,100

Equipment Not Included Above (Elsewhere or of a transient nature)

Augusta Office		50,600
Organized Towns		114,000
Forest Management		28,400
Entomology		25,800
Blister Rust		5,800
Maine Forestry District		226,000
Public Lots		2,100
Maine Mining Bureau		2,000
Totals	$1,877,600	$1,937,500

3.

M.F.D. SUPPRESSION COSTS - 1917-1972
(Emphasis on Labor and Equipment)
(Date from Forestry Department Biennial Reports)

Year	Budget	Suppression Costs		No. of Fires	Acres Burned	Damage	
1917	None	$ 1,265.41		19	147	$ 1,334	
1918	"	7,607.98		58	3,820	7,291	
1919	"	6,402.69		85	4,352	6,305	
1920	"	15,520.54	(1)	118	34,558	143,758	
1921	"	80,120.88		250	56,947	404,555	
1922	"	106,629.69	(2)	164	19,198	106,001	
"		25,801.83	(3)	-	-	-	
1923	"	40,080.61		132	62,407	289,845	
1924	"	61,207.71		158	38,401	101,986	
1925	"	10,802.85		73	2,328	14,058	
"		16,609.70	(4)	-	-	-	
1926	"	33,791.07		83	3,717	34,068	
1927	"	33,031.21		60	9,096	103,649	
1928	"	3,289.63		27	1,562	1,965	
1929	"	5,529.12		90	1,323	11,363	
1930	"	24,087.26		129	11,678	39,316	
1931	"	5,288.85		92	562	1,580	
1932	"	48,399.68		164	36,343	50,731	
1933	"	10,199.69		165	5,299	7,259	
1934	"	54,652.46		165	130,293	385,126	
1935	"	20,506.44		220	14,582	28,001	
1936	"	3,209.72		84	179	13,270	
1937	"	9,536.80		162	1,358	12,191	
1938	"	9,064.70		92	5,210	7,815	
1939	"	7,826.92		128	2,914	15,757	
1940	"	4,965.45		120	523	3,681	
1941	"	41,406.68		157	12,847	82,543	
1942	"	8,617.56		97	1,785	2,853	
1943	"	2,582.60		37	244	4,157	
1944	"	73,661.00		147	12,162	121,773	
1945	"	8,789.98		83	889	4,590	
1946	"	39,260.97		151	3,553	29,482	
1947	"	6,824.33	(5)	167	4,685	20,167	
1948	"	26,430.04	(6)	266	805	33,075	
1949	49,000	73,835.88	(7)	219	16,938	27,618	
1950	49,000	49,877.75		172	6,516	71,334	
1951	52,000	7,690.06		118	503	10,492	
1952	50,000	439,532.97	(8)	301	18,615	535,899	
1953	76,200	215,810.45	(9)	200	4,886	29,506	
1954	65,000	2,933.25		75	453	3,676	
1955	100,000	17,980.80	(10)	168	528	16,362	
1956	100,000	5,931.07		128	552	2,392	
1957	100,000	26,774.87		174	5,745	81,128	
1958	100,000	13,481.79		65	364	3,448	
1959	100,000	34,791.67		137	1,958	26,989	
1960	100,000	139,912.68	(11)	160	1,802	112,357	
1961	100,000	173,452.43	(12)	132	926	17,094	
1962	50,000	13,684.76		111	1,079	31,241	
1963	50,000	56,571.58		207	1,224	39,266	
1964	50,000	30,899.89		202	1,237	10,685	
1965	50,000	37,717.74		211	13,008	-	(13)
1966	52,500	359,114.93	(14)	236	331	-	
1967	52,500	44,573.41		106	980	-	
1968	52,500	101,914.65		211	4,739	-	
1969	52,500	37,483.03		86	309	-	
1970	60,000	24,151.32		158	494	-	
1971	60,000	33,888.79		230	247	-	
1972	60,000	29,807.71		-	-	-	(15)

(1) $8,000 unpaid bills
(2) 1921 deficit
(3) 1922 bills
(4) 1921 and 1924 bills
(5) End of paying bills by Watersheds: St. John, Penobscot, Kennebec, Androscoggin & Machias
(6) Start of paying bills by Supervisor Divisions: Eastern, Northern, Western & Central
(7) Start of budgeting fire suppression bills, includes $9,179.20 bills to be paid in 1950

(8) Includes $108,130.24 to be paid in 1953
(9) Includes $120,986.23 1952 bills paid in 1953
(10) Start of M.F.D. policy of a fixed suppression budget item
(11) End of Calendar Year Budget
(12) Start of Fiscal Year Budget to conform with State procedures
(13) No damage figures as U.S. Forest Service is working on more accurate system of reporting
(14) Contribution from General Fund of $98,463.23 to cover payment of 1965 Centerville Fire bills
(15) Only state total figures shown as Organized Towns and Maine Forestry District have been combined into one entity as result of 1971 internal reorganization

---

4.

STATE OF MAINE

FOREST FIRE REPORT

Rec'd ....
Code No. ....
Checked ....
Mapped ....
Reimb. ....
To Cont. ....

1. Location and Time

Town, City, or Plantation .... Name of Fire .... County .... Weather District

Indicate location of fire, scale 1 in.—3 miles

1. Fire probably started   Date........19........ Hour........ A.M./P.M.
2. Fire discovered   Date........19........ "   .... A.M./P.M.
3. Report received   Date........19........ "   .... A.M./P.M.
4. Crew started for fire   Date........19........ "   .... A.M./P.M.
5. Fire fighting started   Date........19........ "   .... A.M./P.M.
6. Fire controlled   Date........19........ "   .... A.M./P.M.
7. Patrol ceased   Date........19........ "   .... A.M./P.M.

2. Cause (Check One)
- Lightning ☐
- Railroad ☐
- Camp Fires ☐
- Smokers ☐
- Brush or Debris Burning ☐
- Incendiary ☐
- Lumbering ☐
- Miscellaneous ☐

3. Class Responsible (Check One)
- Fishermen ☐
- Hunters ☐
- Tourists ☐
- Berry Pickers ☐
- Farmers ☐
- Construction Crews ☐
- Lumbermen ☐
- Other ☐

4. Resident of (Check One)
- This Town ☐
- State ☐
- Out of State ☐

Check the most likely cause—insert word **probable** if definite proof is lacking

5. Lookout Station reporting fire.................... Date.......... Time.......... A.M./P.M. ☐
6. Who else reported fire............................ Date.......... Time.......... A.M./P.M. ☐
7. Weather.................... Clear, Cloudy, Dry, Moist, Hot, Cold .... Wind .... Direction and Strength
8. Type of Fire: Surface ☐ Crown ☐ Underground ☐
9. Damage and Cost: Acres Burned.......... Damage to Improvements $.......... Suppression Cost $..........
10. Was any fire law violated?.......... Which one?..........
11. What clues did you find?..........
12. Have you started, or do you recommend prosecution?..........
13. Owner of land..........

STATE OF MAINE
FOREST FIRE REPORT

Rec'd MAY 14 1968
Code No. 31.001
Checked eph 5-14-68
Mapped
Reimb. 7.20
To Cont. 5-14-68

1. Location and Time

Municipality: 16 R10 Pub. Lot County: Aroostook Name of Fire: Weisay Brook

1.	Fire probably started	Date 5/12 1968	Hour 3:15	P.M.
2.	Fire discovered	Date " 1968	" 3:25	P.M.
3.	Report received	Date " 1968	" 3:30	P.M.
4.	Crew started for fire	Date " 1968	" 3:35	P.M.
5.	Fire fighting started	Date " 1968	" 3:40	P.M.
6.	Fire controlled	Date " 1968	" 4:45	P.M.
7.	Patrol ceased	Date " 1968	" 7:00	P.M.

Indicate location of fire, scale 1 in.—3 miles

2. Cause (Check One Box and Specific Subdivision)
- ☐ Lightning
- ☐ Campfire
- ☐ Debris Burning
  - Blueberry
  - Dump
  - Incinerator
  - Other
- ☐ Incendiary
- ☐ Machine Use
  - Power Saw
  - Railroad
  - Truck or Tractor
  - Other
- ☒ Smoking
- ☐ Miscellaneous
  - Children
  - Structural Fire
  - Other

3. Activity Class Responsible (Check One)
- Timber Operator ☐
- Recreation ☐
- Other ☒

4. Fire started by Landowner, Agent or Occupant (Check One)
Yes ☐ No ☒

5. Lookout Station reporting fire .... Date .... Time .... A.M./P.M.
6. Who else reported fire DANIEL PELLETIER Date MAY 12 Time 3:25 P.M.
7. Weather HOT Wind S.W. 15 M.P.H.
   Clear, Cloudy, Dry, Moist, Hot, Cold — Direction and Strength
8. Type of Fire: Surface ☒ Crown ☐ Underground ☐
9. Damage and Cost: Acres Burned 150' x 150' Damage to Improvements $ NONE Suppression Cost $ 7.20
10. Was any fire law violated? .... Which one? ....
11. What clues did you find? NO CLUES
12. Have you started, or do you recommend prosecution?
13. Results of court action
14. Owner of land PUB-LOT 16 R10
15. Remarks—Story of Fire

Signature of Town Forest Fire Warden
Address .... Date ....

Form F-6

THIS SIDE FOR STATE PERSONNEL USE ONLY

16. Suppression Information

	Forest Rangers	Labor	Trucks Tractors	Plane	Bulldozers	Food Supplies	Other	Total Costs
Number	4	3				XXXXXXXXXXXX XXXXXXXXXXX		XXXXXXXXXXXXXXXX XXXXXXXXXXXXXXXX
Cost		7.20						$ 7.20

Fire Damage Summary

17. Area Burned   **18.** Damages

Type of Area	Acres
Commercial Forest Land Capable of Producing Crops of Industrial Wood	150' x 50'
Non-Commercial Forest Land Site Incapable of Producing Forest Crops or Withdrawn from Production	
Non-Forest Land Abandoned Fields, Bogs, Etc., Where Fires are Fought to Protect Forest Land	
Total Acres	150' x 50'

Resource	Value
Cut Products	
MBF or Cords @ $ per MBF or Cord	
Recreation	
Wildlife	
Real Property Improvements (Buildings - Mills, etc.)	
Personal Property	
Total Value	

Weather District _____5_____

Spread Index From Nearest Station _____12_____

19. Remarks—Story of Fire FIRE REPORTED BY DANIEL PELLETIER IT WAS TO SMALL FOR ROCKY MT. TO SEE AT THE TIME. IT WAS ON THE EDGE OF OLD GRAVEL PIT. WHERE THEY GET OF HIGHWAY RD. TO DRINK BEER. A CIGG. MUST HAVE BEEN THROUGHED OUT IN DRY GRASS — IT GOT IN THE WOODS — SECOND GROWTH FIR. BROOK WAS HANDY So WE GOT IT STOPED IN A SHORT TIME.

Signature of Forest Ranger _Ronald J Simin_   Date _May 13-62_
Signature of District Forest Ranger _____   Date _____

STATE OF MAINE
FOREST FIRE REPORT

Augusta Office Use Only
Rec'd _____
Checked _____
Reimb. _____
To Cont. _____

1. TOWN	2. COUNTY	3. NAME OF FIRE

	Month	Day	Year 19	Hour	
4. Fire Probably Started			19		am/pm
5. Fire Discovered			19		am/pm
6. Crew Started for Fire			19		am/pm
7. Fire Fighting Started			19		am/pm
8. Fire Controlled			19		am/pm
9. Patrol Ceased			19		am/pm

MAP OF FIRE

Indicate location of fire
scale 1 in. — 3 miles

10. CAUSE (check one cause and specify subdivision)

1. LIGHTNING
2. CAMPFIRE
   (1) Grudge
   (2) Deer burn
   (3) Other
3. DEBRIS BURNING
   (1) Blueberry
   (2) Dump
   (3) Incinerator
   (4) Other
4. INCENDIARY
   (1) Grudge
   (2) Deer burn
   (3) Other
5. MACHINE USE
   (1) Truck or tractor
   (2) Power saw
   (3) Power line or elec. fence
   (4) Other
6. SMOKING
   (1) Camper
   (2) Fisherman or hunter
   (3) Tourist
   (4) Woodsworker
7. RAILROAD
   (1) Exhaust
   (2) Brake shoes
   (3) Other
8. CHILDREN
   (1) Matches
   (2) Campfire
   (3) Smoking
   (4) Other
9. MISCELLANEOUS
   (1) Structure
   (2) Glass
   (3) Other

   (1) Camper
   (2) Fisherman
   (3) Hunter
   (4) Woodsworker

11. CERTAINTY OF CAUSE (check one)
   1. Guess
   2. Probable
   3. Positive

12. FIRE REPORTED BY
   1. Tower _____ Time _____
   2. Aircraft pat. _____ Time _____
   3. Ranger _____ Time _____
   4. Other _____ Time _____

13. TYPE OF FIRE
   [ ] Underground
   [ ] Surface
   [ ] Crown   [ ] Snag

14. MOST COMMON FUEL (check one)
   1. Conifer - mature
   2. Conifer - young
   3. Conifer - slash
   4. Hardwood - mature
   5. Hardwood - young
   6. Hardwood - slash
   7. Brush - shrubs
   8. Grassland

15. LAW VIOLATION
   Which law violated _____
   [ ] No law violated

16. RESULTS OF VIOLATION
   [ ] Convicted
   [ ] Not convicted
   [ ] Paid fire costs
   [ ] Warning
   [ ] Case filed

   Amount paid $ _____
   18. Acres burned _____
   19. Damages _____
   20. Suppression costs _____

17. Owner of land _____
21. REMARKS _____

22. Town Warden Signature _____ 23. Address _____ 24. Date _____

RANGER USE ONLY
Fire Code	Watershed	Map Grid	Crew start	F.F. start	F. control	

OFFICE USE ONLY
Town	cause	C&D	T-F	P. ceased	Law	V-R	Cty	Date

F. start | F. disc. | ... | Snag | ...

Form F-6-71

THIS SIDE FOR STATE PERSONNEL USE ONLY
ACRES AND DAMAGE INFORMATION

Fire Code	Year	25. Man hours to control fire	26. Spread Index	27. Drought Index

28. ACRES BURNED (to the nearest .0 acre, — dotted line indicates decimal point)

TYPE	COMMERCIAL Land Capable of Producing Crops of Industrial Wood	NON-COMMERCIAL Incapable of producing Forest Crops or Withdrawn from Use	NON-FOREST Abandoned fields or Where Fires are Fought to Protect Forests	TOTAL ACRES	Office Use ONLY Type Burn
ACRES					

29. DAMAGES (to the nearest dollar, if any)

	Cut Products	Recreation	Wildlife	Real Property	Personal Property	Total Damages	A+D Card
Damages							2

FIRE COST INFORMATION

Fire Code	Year

30. SUPPRESSION COSTS (actual cost, — dotted line indicates decimal point)

		LABOR		EQUIPMENT		OTHER		TOTAL	Cost Card
Direct Costs	State								
	Vol. Deptn.								
Indirect Costs									
TOTAL									3

31. REMARKS — STORY OF FIRE

_____
_____
_____
_____
_____

SIGNATURE OF FOREST RANGER _____ DATE _____

SIGNATURE OF DISTRICT FOREST RANGER _____ DATE _____

5. AN AFFIDAVIT

I hereby certify that on the second day of June last I was toting goods for the Lincoln Pulp Wood Company from Wissataquoik Lake to Camp on Trout Brook, known as the McCarty camp — The goods being toted to McCarty camps, to and fro by means of a steel shod sled.

On the morning of June second 1915, I left the McCarty Camps on or about 6:30 a.m. in the morning, bound for Wisataquoik Lake. On the way there I stopped for lunch at the storehouse known as Pogie — I arrived at this storehouse about 10 a.m. that same morning — After lunch I went down Pogie mountain via the tote road with my team and Mike McLane, arriving at Wissataquoik Lake about 1:30 p.m. Immediately after loading, (which took about ½ hour) I started to return to McCarty Camps for the night — On going up Pogie mountain toward the storehouse, I had gotten within one half mile from this storehouse, when I commenced to smell fire in the air, or rather smoke from a fire, which led me to believe that timber was in conflagration somewheres.

I hereby certify that on that particular day (June second) I had no matches, pipe or tobacco upon me whatsoever, nor anything that could be used for igniting. Neither to the best of my knowledge did my partner, Mike McLane.

I am a smoking man, however, using about one cut of tobacco per week — on that particular day, however, I had neglected to bring either my pipe or tobacco with me, having left same at home, the McCarty Camps, where upon my arrival that night I saw same on the shelf — I do not use cigarettes or cigars, nor did I on that day. My partner is a smoking man, but on that particular day he did not have his pipe or tobacco either.

I was at a loss to account for the way the fire started, as I was in no way responsible myself for starting same, nor was my partner, Mike McLane — Friction of steel runners against rough stones on the road was the only possibility which suggested itself, and I believe that this is the manner in which it started — the country was very dry at that time, and a very slight spark was indeed sufficient to start a fire with very little effort, and with the dry weather and wind that was blowing, it spread very easily and with rapidity.

I make this statement of my own free will and accord, having asked for my version as to how I thought the fire was started when there was nobody else in the neighborhood, and this, the above recital, is, I believe the manner in which the fire started, and the facts as given above, are to the best of my knowledge and belief true and accurate insofar as I know them. . . . *Interesting sworn statement by Peter Sargent of Patten, Maine, 1915, of innocence in setting or starting a forest fire. This is an extract from a lengthy letter.)*

Appendix VI

1. MAINE FOREST SERVICE
RADIO CALL NUMBERS
1953 Directory

Maine Forestry District

Eastern Division

		Towers
*100 Supervisor	Willard Wight	KCB 429-Cooper Mt.
101 East Machias	Everett A. Grant	603-Musquash Mt.
113 East Machias	Clyde Mattheson	
105 Machias	Macey Armstrong	604-White Cap Mt.
109 Machias	Clarence Dorr	607-Mattamiscontis Mt.
102 Passadumkeag	Emery Lyons	640-Lead Mt.
112 Passadumkeag	George Thompson	641-Schoodic Mt.
....	....	612-Little Russell
106 Pleasant River	Ivan McPheters	163-Portable
103 Pleasant River	Ralph Hartley	
107 St. Croix	Ralph C. Bagley	
108 St. Croix	Harry Noble	
110 St. Croix	Donald Chambers	
104 Union River	Luther G. Davis	
111 Union River	Herman J. Harrington	

Northern Division

*201 Supervisor	Robt. E. Pendleton	KCB 425-Mt. Chase
202 Asst. Supervisor	Glen Tingley	KCB 874-DeBoulie Mt.
207 Aroostook Waters	Harold Weeks	605-Trout Mt.
211 Aroostook Waters	Albert Gagnon	642-Whitney Mt.
209 Aroostook Waters	Harley W. Libbey	643-Ragged Mt.
204 East Branch	Scott Davis	644-No. 9 Mt.
212 East Branch	Floyd Giles	645-Burnt Mt.
206 Katahdin	Emery Grant	161-Portable
213 Katahdin	Clayton G. White	
203 Fish River	Stanley Greenlaw	
214 Fish River	Arnold Shaw	
205 Mattawamkeag	H. Ray Smith	
216 Mattawamkeag	Henry Hunter	
210 Mattawamkeag	William Pratt	
208 Number 9	Earl M. Adams	
215 Number 9	Fred McLean	
....	....	KCB 426-Priestly Mt.
....	....	606-Hardwood Mt.
402 Upper St. John	Chester Goding	639-Rocky Mt.
403 Upper St. John	Arthur Harvey	650-Depot Mt.
404 Musquacook	Lionel Caron	651-Round Mt.
405 Musquacook	Annas F. Bridges	162-Portable
407 Seven Islands	Harold Pelletier	657-Clear Lake
406 Seven Islands	Robert Sinclair	
408 Allagash	Stanley Drake	
409 Allagash	Ronald Simon	
410 Madawaska	Paul Chamberlain	
411 Madawaska	....	

*Two Frequencies

Western Division
Call No.	Location	Name	Frequency
*301	Supervisor	Robert G. Hutton	KCB 428-Squaw Mt.
307	Chesuncook	Oscar Gagnon	KCB 424-Bigelow Mt.
			601-Barren Mt.
303	Dead River	Earle Williams	602-Kibby Mt.
304	Dead River	Bernard A. Tibbits	646-Wadleigh Mt.
313	Dead River	Duluth Wing	
302	Moosehead	John Smith	647-No. 4 Mt.
305	Moose River	Charles Lumbert	648-Old Spec Mt.
			649-Boundary Bald
310	Parlin Pond	Isaac Harris	156-Portable
311	Parlin Pond	Arthur Pillsbury	
309	Rangeley	Kenneth A. Hinkley	
308	Rangeley	Willis Bean	
306	Seboomook	Vaughn Thornton	
312	Seboomook	Roland Peters	

Planes
Call No.	Name	Location
701	Earl Crabb	Augusta
702	Charles Coe	Tramway

Fish and Game Warden Pilots

Call No.	Name	Address	Tel.
705	Wm. H. Turgeon	Augusta	
707	George Later	Greenville	98
708	Malcolm Maheu	Plaisted	Eagle Lake 2589
709	George Townsend	Rangeley	

*Two Frequencies

2. MAINE FORESTRY DISTRICT RADIO CALL AND CAR PLATE NUMBERS
WESTERN REGION 1972

Region Hdq's.

200 Earle Williams
201 Norman Withee
202 John Smith
203 Frank Lavigne
204 George Johnson
280 Hubbard Trefts
281 Maynard Atwood
282 Michael Devine
290 Ronald R. Locke
291
292
293

Aircraft

920 George Johnson (Cessna)
921   "   "   (Beaver)
925   "   "   (Helicopter)

Plane Patrol

970 A Patrol (East Side Moosehead)
971 B Patrol (West Side Moosehead)
972 C Patrol (Moosehead-Rangeley)
973 D Patrol (Arnold Trail)

Moosehead District

210 Vaughn Thornton
211 William Shufelt
212 Ronald Kronholm
213 Robert Merrill
214 Asa Markey
215 Myron Witherell
216 Leroy Knight
217 David Richards
218 Campsite

Seboomook District

220 Galen Cook
221 Charles Howe III
222 Edward Bowden
223 Joel Cyr
224 Stephen Day
225 Austin Sillanpaa
226 Campsite

Arnold Trail District

230 Willis Bean
231 Lewis Prescott
232 Thomas Jones

Parlin Pond District

240 Everett Parsons
241 William Gorham
242 Dale Voter
243 Alan Scamman
244 David Wight

Dead River District

250 Duluth Wing
251 Gilbert Anders
252 Bruce Goodrow
253 Thomas Lemont
254 Warren Bennet
255 Campsite

Rangeley District

260 Terrence Trudel
261 John Hinkley
262 Alfred Roberts
263 Brad Barrett
264 Ralph Clyne
265 Campsite

Utility Vehicles

800-899

Note: Same system of numbering for Eastern and Northern regions

Appendix VII

STATE LAND AGENTS

Year of Appointment	Name	Years of Service	Year of Appointment	Name	Years of Service
1824	James Irish	4	1855	Isaac R. Clark	1
1828	Daniel Rose	2	1856	James Walker	1
1830	Milford P. Norton	1	1857	Noah Barker	3
1831	Daniel Rose	3	1860	B.W. Norris	3
1834	John Hodgdon	4	1863	H. Chapman	1
1838	Elijah L. Hamlin	1	1864	Isaac R. Clark	4
1839	Rufus McIntire	2	1868	Parker P. Burleigh	8
1841	Elijah L. Hamlin	1	1876	Edwin C. Burleigh	8 Mo.
1842	Levi Bradley	5	1877	Fred C. Richards	1
1847	Samuel Cony	3	1877	Edwin C. Burleigh	2
1850	Anson P. Morrill	4	1879	Isaac R. Clark	1
1854	George C. Getchell	1	1880	Cyrus A. Packard	10

STATE LAND AGENTS AND FOREST COMMISSIONERS

Year of Appointment	Name	Years of Service
1891	Cyrus A. Packard	1
1892	Charles E. Oak	9
1901	Edgar E. Ring	10
1911	Frank E. Mace	2
1913	Blaine S. Viles	2
1915	Frank E. Mace	2
1917	Forrest H. Colby	4
1921	Samuel T. Dana	1 Yr. 7 Mo.

FOREST COMMISSIONERS

Year of Appointment	Name	Years of Service
1923	Neil L. Violette	1 Yr. 3 Mo.
1924	Neil L. Violette	11
1935	George H. Gruhn	4 Mo.
1935	Waldo N. Seavey	4
1939	Raymond E. Rendall	8
1948	A. D. Nutting	10 Yr. 5 Mo.
1958	Austin H. Wilkins (Resigned Jan 1, 1973)	14 Yr.

Artistic North Points

3. HONORARY CHIEF FOREST FIRE WARDENS
Lists Selected at Random

CHIEF WARDENS AT LARGE - 1918

Chester W. Alden, Westbrook
Alfred K. Ames, Machias
J.W. Brankley, Portland
H.B. Buck, Bangor
D.A. Crocker, Bangor
Fred A. Gilbert, Bangor
B.W. Howe, Patten
J.C. Hutchinson, Bangor
W.J. Lanigan, Waterville
E.R. Linn, Berlin, N.H.
E.E. Ring, Bangor
B.S. Viles, Augusta

HONORARY CHIEF WARDENS - 1919

Chester W. Alden, Westbrook
Alfred K. Ames, Machias
Guy S. Baker, Ashland
L.P. Barney, Tarratine
R.A. Braman, Portland
H.B. Buck, Bangor
D.A. Crocker, Bangor
Henry Crowell, Skowhegan
W.F. Campbell, Cherryfield
Fred A. Gilbert, Bangor
E.M. Hamlin, Milo
B.W. Howe, Patten
J.C. Hutchinson, Bangor
Frank King, Oquossoc
W.J. Lanigan, Waterville
S.S. Lockyer, Berlin, N.H.
Louis Oakes, Greenville Jct.
J.F. Philippi, Bangor
E.E. Ring, Bangor
H.B. Shepard, Bangor
J.R. Sullivan, Whitneyville
A.B. Sargent, Stratton
Frank Thompson, Skowhegan
B.S. Viles, Augusta
B.M. Winegar, Montreal

HONORARY CHIEF WARDENS - 1921

Chester W. Alden, Westbrook
Alfred K. Ames, Machias
Guy S. Baker, Bangor
L.P. Barney, Tarratine
R.A. Braman, Portland
H.B. Buck, Bangor
W.F. Campbell, Cherryfield
Ira D. Carpenter, Patten
K.M. Clark, Bangor
Forrest H. Colby, Bingham
D.A. Crocker, Bangor
Henry Crowell, Skowhegan
E.B. Draper, Bangor
L.J. Friedman, Great Works
Fred A. Gilbert, Bangor
James Q. Gulnac, Bangor
E.M. Hamlin, Milo
B.W. Howe, Patten
Phil R. Hussey, Bangor
J.C. Hutchinson, Bangor
W.J. Lanigan, Waterville
S.S. Lockyer, Berlin, N.H.
James McNulty, Bangor
Louis Oakes, Greenville Junction
J.F. Philippi, Bangor
E.E. Ring, Bangor
James W. Sewall, Old Town
H.B. Shepard, Bangor
J.R. Sullivan, Whitneyville
A.B. Sargent, Stratton
Frank Thompson, Skowhegan
B.S. Viles, Augusta
B.M. Winegar, Montreal, Canada
Ralph Wing, Flagstaff

HONORARY CHIEF FOREST FIRE WARDENS - 1928

Former Forest Commissioners:
    Forrest H. Colby, Bingham
    Samuel T. Dana, Ann Arbor, Mich.
    Frank E. Mace, Augusta
    Blaine S. Viles, Augusta

Former Deputy Forest Commissioners:
    Charles W. Curtis, Bangor
    Edward A. Mathes, Portland

Former Chief Forest Fire Wardens:
    Alfred K. Ames, Machias
    L.P. Barney, Tarratine
    John M. Brown, Eagle Lake
    H.B. Buck, Bangor
    W.F. Campbell, Cherryfield
    Henry Crowell, Skowhegan
    L.J. Friedman, Great Works
    Fred A. Gilbert, Bangor
    Ora Gilpatrick, Houlton
    Harry E. Hasey, Ashland
    J.C. Hutchinson, Bangor
    W.J. Lanigan, Waterville
    S.S. Lockyer, Berlin, N.H.
    Roy L. Marston, Skowhegan
    Louis Oakes, Greenville Junction
    James M. Pierce, Houlton
    Guy Sedgeley, Stratton
    Frank H. Sterling, Augusta
    J.R. Sullivan, Whitneyville
    B.M. Winegar, Montreal, Canada

HONORARY CHIEF FOREST FIRE WARDENS - 1949

Former Forest Commissioners:
    Samuel T. Dana, Ann Arbor, Mich.
    Raymond E. Rendall, Alfred
    Waldo N. Seavey, Lovell

Former Deputy Forest Commissioners:
    Edward A. Mathes, Portland

Former Supervisor
    Harry G. Tingley

Former Chief Forest Fire Wardens:
    Alfred K. Ames, Machias
    A.P. Belmore, Princeton
    Grover C. Bradford, Bangor
    Ralph L. Brick, Levant
    Edgar I. Carr, Millinocket
    Alex Cormier, Waterville
    Henry Crowell, Skowhegan
    Wm. A. Dubay, Old Town
    Edmund Emery, St. Francis
    E.L. Foss, Milo
    L.J. Friedman, Great Works
    Joseph M. Gagnon, Frenchville
    John Gardner, St. Francis
    V.A. Gilpatrick, Springfield, Ma
    Newman J. Guptill, Milo
    Harry E. Hasey, Bangor
    Errold F. Hilton, Bingham
    William J. Hodgins, Pittsfield
    Arthur L. Holden, Jackman
    Cyril Jandreau, St. Francis
    Herbert Johnston, Rockwood
    Anton R. Jordan, Aurora
    Frank C. King, Oquossoc
    Emil J. Leavitt, Old Town
    Harry McReavy, Whitneyville
    John E. Mitchell, Patten
    C.C. Murphy, Rangeley
    Louis Oakes, Greenville Junction
    Thos. Perrow, Millinocket
    James M. Pierce, Houlton
    Richard Pierce, Gardiner
    George P. Ryan, E Millinocket
    Ralph Sterling, Caratunk
    John V. Wing, Eustis
    Harold W. York, Rangeley

HONORARY CHIEF WARDENS - 1973

Albert Baker, T15 R15
  (Retired and moved, 1973)
Maurice Bartlett, Ashland
Grover C. Bradford, Bangor
  (deceased)
Annas Bridges, Masardis
Edmund J. Emery, Sheridan
George Faulkner, Ellsworth
Emery B. Grant, Millinocket
Everett A. Grant, Dennysville
William J. Hodgins, Pittsfield
Harold J. Pelletier, Caribou
James M. Pierce, Houlton
John G. Sinclair, Bangor
Helen Taylor, Farmington

4.   CHIEF WARDENS IN CHARGE OF RAILROAD PATROL - 1914

R.J. Anderson (Brownville), Territory in Maine Forestry District from Mattawamkeag to Boundary on Canadian Pacific Railroad.
H.B. Stimson (Houlton), Territory in Maine Forestry District North of Oakfield Jct., on Bangor & Aroostook Railroad.
F.H. Gould (Milo), Territory in Maine Forestry District from Oakfield Jct., to Milo Jct., on Bangor & Aroostook Railroad.
Forrest H. Colby (Bingham), Territory in Maine Forestry District on Somerset Division of the Maine Central Railroad.
C.C. Murphy (Rangeley), Territory in Maine Forestry District on Oquossoc Division of the Maine Central Railroad and also Rangeley Division of the Sandy River and Rangeley Lakes Railroad.
A.R. Henderson (Kingfield), Territory in Maine Forestry District on Bigelow Division of the Sandy River and Rangeley Lakes Railroad.

1915

Chas. Powers (Brownville), Territory in Maine Forestry District from Mattawamkeag to Boundary on Canadian Pacific Railroad.
H.B. Stimson (Houlton), Territory in Maine Forestry District North of Oakfield Jct., on Bangor & Aroostook Railroad.
Leroy Haley (Glenburn), Territory in Maine Forestry District from Oakfield Jct., to Milo Jct., on Bangor & Aroostook Railroad.
Albert F. Webster (Bingham), Territory in Maine Forestry District on Somerset Division of the Maine Central Railroad.
C.C. Murphy (Rangeley), Territory in Maine Forestry District on Oquossoc Division of the Maine Central Railroad and also Rangeley Division of the Sandy River and Rangeley Lakes Railroad.
Philander Butts (Kingfield), Territory in Maine Forestry District on Bigelow Division of the Sandy River and Rangeley Lakes Railroad.
Leon C. Irish (Haynesville), Territory in Maine Forestry District from Kingman to Wytopitlock, Maine Central Railroad.
Fred Houghton (Topsfield), Territory in Maine Forestry District from Forest to Vanceboro, Maine Central Railroad.
Fred S. Bunker (Franklin), Territory in Maine Forestry District from Washington Jct. to Unionville, Maine Central Railroad.

1916

Chas. Powers (Brownville Jct.), Right of way of the Canadian Pacific Railroad in Maine.
A.O. Holden (Oakfield), Right of way of the Bangor and Aroostook R.R. north of Oakfield Jct.
Leroy Haley (Glenburn), Right of way of the Bangor and Aroostook R.R. south of Oakfield Jct.
Frank Hilton (Bingham), Right of way of the Maine Central R.R., Somerset division.
C.C. Murphy (Rangeley), Right of way of the Maine Central R.R., Oquossoc division.
Philander Butts (Kingfield), Right of way of the Sandy River and Rangeley Lakes R.R.
Leon C. Irish (Haynesville), Territory in Maine Forestry District from Kingman to Wytopitlock, Maine Central Railroad.
Fred Houghton (Topsfield), Territory in Maine Forestry District from Forest to Vanceboro, Maine Central Railroad.
Fred S. Bunker (Franklin), Territory in Maine Forestry District from Washington Jct. to Unionville, Maine Central Railroad.

1917

W. Garland (Brownville Jct.), Right of way of the Canadian Pacific Railway in Maine.
A.R. Henderson (Kingfield), Right of way of the Sandy River and Rangeley Lakes railroad.
Frank W. Hilton (Bingham), Right of way of the Maine Central R.R. Somerset division.
F.H. Gould (Milo), Right of way of the Bangor and Aroostook R.R.
C.C. Murphy (Rangeley), Right of way of the Maine Central R.R. Oquossoc division.
S.C. Cummings (Haynesville), Territory in Maine Forestry District from Kingman to Wytopitlock, Maine Central Railroad.
J.J. Kneeland (Topsfield), Territory in Maine Forestry District from Forest to Vanceboro, Maine Central Railroad.
Fred S. Bunker (Franklin), Territory in Maine Forestry District from Washington Jct. to Unionville, Maine Central Railroad.

1921

W. Garland (Brownville Jct.), Right of way of the Canadian Pacific Railroad in Maine.
A.R. Henderson (Kingfield), Right of way of the Sandy River and Rangeley Lakes Railroad.
S.C. Cummings (Haynesville), Right of way of the Maine Central Railroad from Kingman to Bancroft.
J.J. Kneeland (Topsfield), Right of way of the Maine Central from Forest to Vanceboro.
Fred S. Bunker (Franklin), Right of way of the Maine Central from Washington Junction to Cherryfield.
C.C. Murphy (Rangeley), Right of way of the Maine Central Railroad, Oquossoc Division.
John Comber (Caratunk), Right of way of the Maine Central Railroad, Somerset Division.
B.S. Archibald (Derby), Searsport to Brownville, Derby to Greenville, Brownville to Iron Works, right of way of the Bangor and Arostook Railroad.
Chas. Tweedie (Millinocket), Old Town to South Lagrange, South Lagrange to West Seboois via Medford, Brownville to Patten Junction, Millinocket to East Millinocket, Patten Junction to Patten on right of way of the Bangor and Aroostook Railroad.
George Dinsmore (Houlton), Patten Junction to Phair, Phair to Fort Fairfield, Ashland Junction to Squa Pan, on right of way of Bangor and Aroostook Railroad.
R.J. Berryman (Fort Kent), Squa Pan to Stockholm, Mapleton to Presque Isle, Phair to Van Buren, Caribou to Limestone, on right of way of Bangor and Aroostook Railroad.
Frank Wood (Fort Kent), Squa Pan to Fort Kent, Van Buren to St. Francis, on right of way of Bangor and Aroostook Railroad.

5.

DEPUTIES

BANGOR & AROOSTOOK RAILROAD
Glenburn to South Lagrange

Name	Address	Section No.
W.H. Chaples, Foreman	Hudson	111
C.G. Chaples, Foreman	Hudson	112
F.D. Messer, Foreman	South Lagrange	113

Medford to Packards

E.E. Chaples, Foreman	Medford	116
B.C. Forrest, Foreman	Rand Cove	117
Frank McGrath, Foreman	Rand Cove	118

Monson Jct. to Greenville

L.S. Mitchell, Foreman	Monson Junction	14
Lyman E. Davidson, Foreman	Blanchard	15
W.W. Fitzsimmons, Foreman	Shirley	16
F.S. Wyman, Foreman	Greenville	17

Brownville Jct. to Iron Works

Edw. Fossman, Foreman	Brownville Jct.	8

Brownville to Dyer Brook

J.B. Porter, Foreman	Brownville	22
W.J. Forrest, Foreman	Schoodic	23
Edw. Helstrom, Foreman	Schoodic	24
J.W. Lindsay, Foreman	West Seboois	25
D.E. McGrath, Foreman	West Seboois	26
Elmer Pease, Foreman	Norcross	27
E.E. Trafton, Foreman	Millinocket	28
J.A. Gaskin, Foreman	Millinocket	29
R.D. Porter, Foreman	Grindstone	30
Curtis R. McKenney, Foreman	Grindstone	31
Charles Box, Foreman	Stacyville	32
Joseph Giddings, Foreman	Sherman	33
David McClarie, Foreman	Crystal	34
Edw. Appleby, Foreman	Island Falls	35
M.M. McDonald, Foreman	Dyer Brook	36

Smyrna Mills to Eagle Lake

Avon Chambers, Foreman	Smyrna Mills	60
Hanford Foster, Foreman	Howe Brook	61
Wallace Porter, Foreman	Howe Brook	62
Chas. Ewings, Foreman	Griswold	63
Harry Ewings, Foreman	Griswold	64
Robt. Whitehouse, Foreman	Masardis	65
Joseph Chambers, Foreman	Squa Pan	66
Geo. Waddington, Foreman	Ashland	67
Joseph Caron, Foreman	Sheridan	68
John A. Boone, Foreman	Portage	69
Wm. Gilpatrick, Foreman	Portage	70
E.G. Bartlett, Foreman	Winterville	71
Jos. Levesque, Foreman	Winterville	72
A.A. Frennette, Foreman	Eagle Lake	73

Monticello to Bridgewater

Henry Faulkner, Foreman	Monticello	43
Frank Everett, Foreman	Bridgewater	44

New Sweden to Van Buren

Wilmot Wibberly, Foreman	New Sweden	53
Harry Dixon, Foreman	Stockholm	54
Abram Grant, Foreman	Stockholm	129
Michael Levesque, Foreman	Van Buren	130

Squa Pan to Mapleton

D.B. McDonald, Foreman	Squa Pan	121
Eber Barrows, Foreman	Mapleton	122
O.H. Pettengill, Foreman	Mapleton	123

Perham to Stockholm

Robert Glew, Jr., Foreman	Perham	126
Frank Wright, Foreman	Perham	127
A. Medley Glew, Foreman	Stockholm	128

6. CANADIAN PACIFIC RAILROAD CO.
Mattawamkeag Subdivision

J. Robichaud, Foreman	Mattawamkeag	2
E.A. Philbrook, Foreman	Chester	3
R. Archer, Foreman	Woodard	4
W.E. Nason, Foreman	Seboois	5
R.B. Brown, Foreman	Hardy Pond	6
J. Newman, Foreman	Lake View	7
V. Newman, Foreman	Brownville Jct.	8

Moosehead Subdivision

J. Sullivan, Foreman	Brownville Jct.	1
J. Meulendyke, Foreman	Williamsburg	2
N. Laroche, Foreman	Barnard	3
I. Philbrook, Foreman	Onawa	4
W.L. Brown, Foreman	Camp 12	5
A. Badeau, Foreman	Morkill	6
J. Dube, Foreman	Greenville	7
J. Conley, Foreman	Squaw Brook	8
P. James, Foreman	Moosehead	9
A. Girard, Foreman	Tarratine	10
A. Dubois, Foreman	Brassua	11
P. Ferland, Foreman	Mackamp	12
O. Maillet, Foreman	Long Pond	13
C.J. Achey, Foreman	Jackman	14
P. Nadeau, Foreman	Attean	15
M. Thibodeau, Foreman	Holeb	16
E. Plante, Foreman	Keough	17
E. Gagnon, Foreman	Lowelltown	18
D. Rioux, Foreman	Boundary	19

7. MAINE CENTRAL RAILROAD COMPANY
Kingman to Bancroft

Joseph E. Meaghear, Foreman	Kingman	104
Martin Faraday, Foreman	Bancroft	107

Forest to Vanceboro

Edward Grass, Foreman	Forest	110
William Trask, Foreman	Lambert Lake	111
Albert Russell, Foreman	Vanceboro	112

Washington to Unionville

Albert McLaughlin, Foreman	Washington Jct.	171
Edward P. Garbett, Foreman	Franklin	172
Elbridge G. Chandler, Foreman	Unionville	174

Houghton to Kennebec

Albert C. Hodsdon, Foreman	Houghton	217
Winfield S. Rose, Foreman	Summit	218
Algernon L. Eastman, Foreman	Bemis	219
George Storer, Foreman	Oquossoc	221
Chas. N. Jacques, Foreman	Oquossoc	222
Chas. H. Carey, Foreman	Kennebago	223

Bingham to Kineo

Harley A. Strout, Foreman	Bingham	249
William Chamberlain, Foreman	Deadwater	251
Romeldo O'Neal, Foreman	Troutdale	252
Wm. M. Otis, Foreman	Lake Moxie	254
Lee Heath, Foreman	Lake Moxie	256
John Smedberg, Foreman	Tarratine	257
Fred D. Kennedy, Foreman	Rockwood	258
Malon Tracy, Foreman	Lake Austin Bald Mt. No. 1	
Stephen Holt, Foreman	Lake Austin Bald Mt. No. 2	
John Hutchins, Foreman	Oakland	Kineo Branch
C.B. Lord, Foreman	Oakland	Kineo Branch

8. SANDY RIVER & RANGELEY LAKES RAILROAD

Farmington to Rangeley

John Tardy	Farmington	1
David Richardson	Strong	2
W.W. Sellinger	Phillips	3
V.H. Huntington	Phillips	4
M.F. Johnson	Phillips, R.F.D.	5
J.R. Wyman	Redington	6
E.F. McCourt	Redington	7
Robert Nile	Rangeley	8

Strong to Bigelow

S.L. Kennedy	Strong	9
M.M. Baker	Salem	10
A.L. Stevens	Kingfield	11
Fred Parsons	Kingfield	12
N.I. McCollar	Bigelow	13

Madrid Branch No. 6

I.L. Haley	Phillips, R.F.D.	14

Grey Farm Branch

C.A. Plummer	Phillips, R.F.D.	17

Perham Branch

A.C. Corson	Phillips, R.F.D.	15

Eustis Branch

W.E. Billington	Dallas, Maine	16

Langtown Branch

V.J. Batchelder	Dallas, Maine	18

9. WARDENS APPOINTED IN 1909 NOW LIVING *

Chief Warden	Louis Oakes, Greenville
Deputy Wardens	Blin W. Page, Skowhegan
	William McNally, Portage
	Leon Orcutt, Ashland
	Alphonse Blanchette, St. Pamphile, Quebec
	Guy Johnson, Baltimore, Maryland
	Alton Carl, Bingham
	Robie Howes, Bingham
	Ray Viles, Portland
	John Comber, The Forks
	J. A. Durgin, The Forks
	Freeland D. Abbott, Houghton
	George E. King, Jr., Bethel
	Z. L. Harvey, Florida
	Lloyd Houghton, Bangor

* At time of 50th M.F.D. anniversary, 1959.

1959 REMINISCENCES BY TWO 1909 WARDENS

An interview with Chief Warden Louis Oakes who is today 89 years old and active in forest conservation:

"The best fire control is to stop fires before they get started."

"We must use a little of the past to help and guide us in the present and possibly extend to the future."

"Men used on a fire are like materials and must be handled properly."

Mr. Oakes, during the interview, stated that the only fire tools used were the axe, pail, shovel, and cut boughs. The food came from lumber camps and the men were well fed. Fire fighters were native woodsmen. Causes of fires were largely by pipe heels (cigarettes not known in those days), lightning, and land clearing. Sap carrying yokes to carry pails of water were tried out but proved too cumbersome for woods travel. In 1913, a man named Dorre, of Dover-Foxcroft, designed a plunger type pump for pumping water from pails onto the fire.

Lloyd E. Houghton:—"As one of the few living deputy wardens who over 50 years ago was employed by the Maine Forest Service soon after it was organized, it has been my good fortune to see this organization develop from rather a crude beginning into one of the most efficient forest protection agencies in the United States.

"The fire fighting units and the men who have charge of them are surpassed by none.

"Much headway is being made in insect control by trained men.

"It is very encouraging to see the close cooperation between the timber landowners in the Maine Forestry District and the Maine Forest Service in an effort to protect and perpetuate Maine's greatest asset."

10. LAWS REGARDING APPOINTMENTS

County commissioners of each county in which there are unorganized places *shall annually appoint,* when they deem it necessary, such number of fire wardens as they deem necessary not exceeding ten, for all such unorganized places in any county, whose duties and powers shall be the same with respect to such unorganized places as those of the fire wardens of towns, and they shall also have the same authority to call out citizens of the county to aid them in extinguishing fires, that town wardens have to call out citizens of the town. The compensation of such fire wardens shall be paid by the county, and the compensation of persons called upon by them as aforesaid, to render aid shall be the same as that provided in the case of towns and shall be paid one-half by the county and one-half by the owners of the lands on which said fires occur.

Ref: Chapter 100, Section 4, Public Laws 1891 — Maine

The said commissioner *shall appoint* in and for each of said districts so established, a chief forest fire warden, and he shall also appoint within such districts such number of deputy forest fire wardens as in his judgment may be required to carry out the provisions of this act, assigning to each of the latter the territory over and within which he shall have jurisdiction. All chief and deputy forest fire wardens, so appointed, shall hold the office during the pleasure of said commissioner, be sworn to the faithful discharge of their duties by any officer authorized to administer oaths, and a certificate thereof shall be returned to the office of such commissioner.

Ref: Chapter 193, Section 8, Public Laws 1909

The commissioner shall appoint, subject to the Personnel Law, a Deputy Forest Commissioner, a State Entomologist, foresters, officers, forest rangers and other expert and clerical assistants as may be necessary. All forest rangers shall be sworn to the faithful discharge of their duties and all persons employed by him shall not be concerned directly or indirectly in the purchase of state lands, nor of timber or grass growing or cut thereon except in their official capacity. They may be allowed actual necessary expenses of travel. Whenever the term "commissioner" is used in chapters 201 to 215 it shall include his agents and representatives.

Ref: Title 12, M.R.S.A., Sub-Chapter I, Section 521, 1964

The commissioner may appoint general deputy wardens as an adjunct to the personnel regularly employed in the forest fire control program. They shall aid in forest fire prevention and shall take immediate action to control any unauthorized forest fires, employ assistance when required and notify the nearest forest ranger or town forest fire warden with dispatch. Such general deputy wardens and those they employ may receive the prevailing local fire fighting wages for the period so engaged.

Ref: Title 12, M.R.S.A., Sub-Chapter II, Section 523, 1964

11. CHAPTER 63
STATE EMPLOYEES APPEALS BOARD

New Sections
751. State Employees Appeals Board.
752. Mediation authority.
753. Procedure for settlement.

§ 751. State Employees Appeals Board

There is established an impartial board of arbitration to be known as the State Employees Appeals Board which shall consist of 3 members to be appointed by the Governor, with the advice and consent of the Executive Council, from persons not employed by the State of Maine and who have established background positively indicating a capacity to mediate grievances between management and labor, one of whom shall be an attorney admitted to practice law in this State. Of those members first appointed, one shall be appointed for a term of one year, one for 2 years and one for 3 years. Their successors shall be appointed for 3 years. The compensation of the members of the board shall be fixed by the Governor and Council. The members of the board shall receive their necessary expenses.

The board shall:

1. *Administration.* Administer this chapter. In exercising its administration, the board may promulgate operating policies, establish organizational and operational procedures, and exercise general supervision. The board shall employ, subject to the Personnel Law, such assistants as may be necessary to carry out the purposes of this chapter.

2. *Rules and regulations.* Promulgate such rules and regulations as are necessary to effectuate the purposes of this chapter.

3. *Report.* Report biennially to the Governor and Legislature facts and recommendations relating to the administration and needs of the board.

1968, c. 539, § 1.

Amendments:
—1968. Chapter new.

§ 752. Mediation authority

The board shall have the authority to mediate the final settlement of all grievances and disputes between individual state employees, both classified and unclassified, and their respective state agencies, except in matters of classification and compensation. All complaints between a state employee and the state agency by which he is employed shall be made and heard in the manner provided by this chapter for the mediation and settlement of such complaints. During the procedure for settlement, an employee may be represented at each step by his designated representative. The decision of the board shall be final and binding upon the state agency and state employees involved in the dispute, and shall supersede any prior action taken by the state agency with reference to the employment and working conditions of such employees.

1968, c. 539, § 1.

Amendments:
—1968. Enacted this section.

§ 753. Procedure for settlement

A grievance or dispute between a state employee and the agency of the State by whom he is employed shall be entertained by the board upon the application of the employee, providing there shall have been compliance with the following requirements;

1. *Adjust dispute.* That the employee aggrieved by the dispute and the employee or his representative, or both, shall have attempted to adjust the dispute with the employee's immediate supervisor.

2. *Grievance in writing.* If the employee is dissatisfied with the oral decision of his immediate supervisor, he may present the grievance to his supervisor again, this time in written form. The supervisor is then required to make his decision in writing and present it to the employee within 3 working days.

3. *Appeal to department head.* If the employee is dissatisfied with the supervisor's written decision, he then may appeal, in writing, to the department head. Within 3 working days, the employee shall receive, in writing, the department head's decision.

4. *Meeting.* If the employee is dissatisfied with the department head's written decision, the department head shall meet with the employee or his representative, or both, and attempt to adjust the dispute. At least one day prior to such meeting, the employee's representative, if any, shall have access to the work location of the employee involved during the working hours for the purpose of investigating the causes of the grievance.

5. *Appeal to Director of Personnel.* If the classified employee is dissatisfied with the decision, following a meeting with the department head, he shall appeal to the Director of Personnel who shall, within 6 working days, reply in writing, to the aggrieved employee and the department head involved in his decision, based on the state's personnel law and rules.

6. *Submission to board.* In the event the grievance shall not have been satisfactorily adjusted within 2 weeks under subsections 1 to 5, the dispute shall be submitted to the board which shall investigate the matters in controversy, shall hear all interested persons who come before it, and make a written decision thereof, which shall be binding on the parties involved. The board's written decision shall be issued within 30 days after the dispute is submitted, unless both parties agree that an extension of this time limit should be allowed.

1968, c. 539, § 1.

Amendments:

—1968. Enacted this section.

---

12. CHAPTER 147

AN ACT Placing All Unclassified State Forestry Department Employees in the Classified System.

Be it enacted by the People of the State of Maine, as follows:

State Forestry Department employees presently unclassified to be placed in the state classified system. The State Personnel Board is directed to amend the compensation plan of the State of Maine, as provided in the Revised Statutes, Title 5, section 634, so that the unclassified employees in all classifications of the State Forestry Department shall be appropriately classified and placed on the same type of step salary range schedule as are other classified employees in the state service employed under the Personnel Law and approved by the Personnel Board.

Employees thereafter shall be accorded the benefits as provided for in the Revised Statutes, Title 5, section 634.

Effective October 1, 1969

---

13. MEMO TO: Maine Forestry District Personnel
FROM: Austin H. Wilkins, Forest Commissioner
SUBJECT: Salary Adjustment

Clearance has been obtained to provide pay to rangers in the Maine Forestry District at the same rate as equivalent positions are paid in the Organized Towns Division. This is a one range increase to become effective September 1. (First check increase will be September 11, 1969.) Pay ranges effected are listed below:

Forest Ranger II (Unit Ranger in O.T. & M.F.D.)
    Range 12, Step A – E, $103.50 – $123.50

Forest Ranger III (Assistant District Ranger in M.F.D.)
    Range 15, Step A – E, $118.00 – $141.50

Forest Ranger IV (District Ranger in O.T. & M.F.D.)
    Range 18, Step A – E, $135.00 – $162.50

Review of the small differences between equivalent jobs in the Maine Forestry District and Organized Towns and consideration of the wider differences within any given classification lead to this decision.

August 27, 1969

14. *One of the earlier certificates of appointment forms. Date unknown*

State of Maine.

County of .... ss.

On the .... day of .... A. D., 19

personally appeared .... of ....

in said County of .... and took and subscribed the

Oaths prescribed by the Constitution of this State and a law of

the United States, to qualify him to discharge and execute the

duties of the office of .... Forest Fire Warden within

and for the State of Maine, to which he was appointed and com-

missioned on the .... day of .... 19 .

Before me,

.... { Authorized to Administer Oaths

Sir---I have the honor herewith to transmit the evidence of

the qualification of ....

as a ....

and am, Sir,

Very respectfully your obedient servant,

....

N. B. *This Certificate should be immediately forwarded to the Forest Commissioner, Augusta, Maine.*

1918

State of Maine.

To all who shall see these presents.

GREETING:

Be it Known, That I, Forrest H. Colby, Forest Commissioner, reposing confidence in the ability, discretion and integrity of John E. Mitchell of Patten in the county of Penobscot do appoint him a **Chief Forest Fire Warden**, at a compensation of $3.00 per day, and actual expenses, to act in the following territory:

Twps. 6, R. 6; 4,5,6,7,8, R. 7; 4,5,6,7, R. 8; 4,5,6,7, R. 9; 5,6,7, R. 10; 5, R. 11.

The said Chief Warden shall hold office, when qualified, for balance of year, unless sooner removed by the Commissioner, and perform such duties, at such times, and under such rules and regulations as he may from time to time prescribe.

In testimony whereof, I the said Forest Commissioner, have hereunto set my hand, at Augusta, this fourth day of April in the year of our Lord one thousand nine hundred and eighteen and of the Independence of the United States of America the one hundred and forth second.

Forrest H. Colby
Forest Commissioner.

STATE OF MAINE

Patten, Maine, May 1, 1918.

Personally appeared John E. Mitchell, and took and subscribed the oaths prescribed by the Constitution of this State and a law of the United States, to qualify him to execute the trust reposed in him by the within Commission.

Before me,

Raymond D. Gardner.

{ Authorized to Administer Oaths

Maine Forest Service

1925

To all who shall see these presents.

GREETING:

Be it Known, That I, Neil L. Violette, Forest Commissioner, reposing confidence in the ability, discretion, and integrity of John E. Mitchell of Patten, in the county of Penobscot, do appoint him a Chief Forest Fire Warden, at a compensation of $ 4.00 per day, and an allowance of $ .75 per day for subsistence, or actual expenses, to act in the following territory:

Townships 6, 7, R. 6; 4, 5, 6, 7, R. 7; 4, 5, 6, 7, R. 8; 4, 5, 6, 7, So. ½ 8, R. 9; 3 (E. ½), 4 (E. ½), 5, 6, 7, R. 10, W. E. L. S.; No. ½ Stacyville Pl.

East Branch District

The said Warden shall hold office, when qualified, for balance of year, unless sooner removed by the Commissioner, and perform such duties, at such times, and under such rules and regulations as the latter may from time to time prescribe.

In Testimony Whereof, I, the said Forest Commissioner, have hereunto set my hand and seal, at Augusta, this nineteenth day of March in the year of our Lord one thousand nine hundred and twenty-five, and of the Government of the State of Maine the one hundred and sixth.

Neil L. Violette
Forest Commissioner

State of Maine

COUNTY OF Penobscot ss. April 11 1925

Personally appeared John E. Mitchell before me and took oath that he would faithfully discharge the duties devolving upon him as Chief Forest Fire Warden within the territory within which he is to act under the foregoing appointment.

Eunice J. Mitchell
Justice of the Peace

STATE OF MAINE
Maine Forest Service
Certificate of Appointment

This is to certify that _____
of _____ in the County of _____
has been appointed a _____

Forest Fire Warden to act in the following Territory:

_____
_____

until _____ unless sooner removed
Date _____
(OVER)   Forest Commissioner

KEEP MAINE GREEN

Return this signed and sworn to section to Augusta

STATE OF MAINE

County of _____ ss. Date _____
personally appeared _____ before
me and made oath that he would faithfully discharge the duties
as _____ Forest Fire Warden within
the territory in which he is to act under the foregoing appointment.

_____
Justice of Peace or Notary Public

PREVENT FOREST FIRES

STATE OF MAINE — FORESTRY DEPARTMENT
CERTIFICATE OF APPOINTMENT

This is to certify that _____
of _____ County of _____
has been appointed _____
authorized territory _____
for period _____ unless sooner removed.
Date _____ Forest Commissioner _____
Signed _____

County of _____ Date _____
Personally appeared _____ before me and made oath
that he would faithfully discharge the duties of _____
within above-mentioned territory.

_____
Justice of Peace or Notary Public

Form F-23   Complete both copies — Return carbon copy to Augusta

---

STATE OF MAINE
FORESTRY DEPARTMENT
EMPLOYEE IDENTIFICATION

EMPLOYEE		
DIVISION		DATE ISSUED
SOCIAL SECURITY NUMBER	HEIGHT	WEIGHT	EYES

THIS CARD MUST BE SURRENDERED
TO THE FORESTRY DEPARTMENT UPON
TERMINATION OF EMPLOYMENT

_____
FOREST COMMISSIONER

EMPLOYEE'S SIGNATURE

State of Maine
FORESTRY DEPARTMENT

In accordance with Title 12, Chapter 215, M.R.S.A. _____

_____
of _____ in the County of _____ is hereby appointed as

Forest Ranger
For the State of Maine

until relieved of duty.

Given under my hand at Augusta, this _____ day of _____ 19 _____

STATE OF MAINE
_____
Forest Commissioner

_____ ss.   _____ 19 _____

Then personally appeared the above named _____
and made oath that he would faithfully and impartially perform the duties required of him by this Commission.

Most Current Form Used

Before me _____
Justice of the Peace

295

15.

The Annual Meeting
of the
Timberland Owners
of the
Penobscot Waters

Will be held in the Chamber of Commerce Rooms
Bangor, on the
13th day of March, 1934 at 2.00 o'clock P. M.

The purpose of the meeting is to talk with the Forest Commissioner, the appointment of fire wardens for the ensuing year, and any other matters that will be of interest. You are cordially invited to attend.

NEIL L. VIOLETTE,
Forest Commissioner

March 3, 1934

The Annual Meeting
of the
Timberland Owners
of the
Penobscot Waters

Will be held in the Chamber of Commerce Rooms
Bangor, on the
10th day of March, 1936 at 2.00 o'clock P. M.

The purpose of the meeting is to talk with the Forest Commissioner, the appointment of fire wardens for the ensuing year, and any other matters that will be of interest. You are cordially invited to attend.

WALDO N. SEAVEY
Forest Commissioner

March 2, 1936

The Annual Meetings
of the
Timberland Owners
For the year 1940 will be held as follows:

Kennebec and Androscoggin Waters at the office of the Forest Commissioner, State House annex, Augusta, on the 6th day of March, at 10 o'clock A. M.

Penobscot Waters at the Chamber of Commerce Rooms, Bangor, on the 12th day of March, at 2 o'clock P. M.

Washington and Hancock Counties at the office of the Eastern Pulpwood Co., Calais, on the 14th day of March, at 10 o'clock A. M.

St. John Waters at the Court House (Grand Jury Room), Houlton, on the 20th day of March, at 10 o'clock A. M.

These meetings will be held to discuss matters of mutual coöperation between the land owners and the Maine Forest Service.

RAYMOND E. RENDALL
Forest Commissioner

February 24, 1940

16. GUARDIANS OF THE FORESTS OF MAINE

In the spring of 1919 the Forestry Department instituted a new project; namely, the organization of the Guardians of the Forests of Maine. The first meeting was held at Augusta April 8th and 9th and attended by over forty chief forest fire-wardens and representatives of the timberland owners. The program of this meeting was as follows:

TUESDAY FORENOON

Commissioner and Deputy Commissioner met the Wardens at 9:36 and 9:50 trains.

10:00 o'clock
Assignment of rooms at the Augusta House

10:30 o'clock
Meeting in Senate Chamber at State House
Calling of the roll

10:45 o'clock
Address of welcome by the Commissioner

11:15 o'clock
Question Box

12:00 Noon
Luncheon at the Augusta House

TUESDAY AFTERNOON

2:00 o'clock
Address, Maine's Forests — HON. BLAINES S. VILES, *Former Forest Commissioner*

2:30 o'clock
First Aid Work — MAJOR BIAL F. BRADBURY

3:00 o'clock
Address on Fish and Game — COMMISSIONER WILLIS E. PARSONS

4:00 o'clock
Moving Pictures at Colonial Theatre

TUESDAY EVENING

7:00 o'clock
Banquet at Augusta House
Music by Merrill's Orchestra
Address, "My trip across as Manager of the New England Sawmill Units" — E. C. HIRST, *State Forester of New Hampshire*

WEDNESDAY FORENOON
7:30 o'clock
A Real Woods Breakfast, Augusta House
8:30 o'clock
The Commissioner and Deputy Commissioner, in the offices at the State House, conferred and advised with the Wardens in regard to the work for each Warden's own territory
11:00 o'clock
Meeting in the Senate Chamber for final talk and final adjournment
12:00 Noon
Farewell luncheon at Augusta House.

REVISED PROGRAM – WINTER STAFF MEETING
DEPARTMENT OF CONSERVATION, BUREAU OF FORESTRY
December 11, 12, 13, 1973 – Augusta Civic Center

TUESDAY, DECEMBER 11

1:00–5:00 P.M.　Functional Division Meetings
　Forest Fire Control .... Kennebec Room, Al Willis
　Forest Management ......:... Androscoggin Room, Robert Dinneen
　Entomology .... Aroostook Room, Robley Nash

Evening Open.　Four hospitality rooms in the Howard Johnson will be available for informal gathering. Exhibits will be available for all to see.

WEDNESDAY, DECEMBER 12

7:00 A.M.　Buffet breakfast or dine from menu at Howard Johnson (Tickets for buffets must be picked up at desk in the lobby when registering for room)
8:00 A.M.　General Session ...... Cushnoc Auditorium, Fred E. Holt
8:00–8:15 A.M.　Open Remarks .... Fred E. Holt
8:15 A.M.　The New Department .... Dr. Donaldson Koons
　　　　　　of Conservation .... Dr. Donaldson Koons
9:15 A.M.　Public Lands Committee .. Senator Harrison Richardson
10:00 A.M.　Coffee and Danish
10:20 A.M.　Information and Education .... Walter Gooley
10:45 A.M.　The North Maine Woods .... Al Leighton,
　　　　　　　　　　　　　　　　　Seven Islands Land Company
11:40 A.M.　Buffet Luncheon, Piscataquis/Sagadahoc Room
　　　　　　Luncheon, Awards, Presentations .... Fred E. Holt
　　　　　　"Impressions of Europe" .... Austin H. Wilkins
1:30 P.M.　Environmental Laws .... Henry Warren,
　　　　　　　　　　　　　　　Dept. Environmental Protection
2:30 P.M.　Maine Management and Cost Survey ...... Fred E. Holt
　　　　　　U.S.F.S. General Program Review .... Temple Bowen
3:15 P.M.　Coffee Break

3:35 P.M.	Review of 1973 Spruce Budworm Spray	Robley Nash
	Project and What's in the Future	Robley Nash
3:50 P.M.	Spruce Budworm Salvage	Joseph Lupsha
4:15 P.M.	Interesting Pests	Louis Lipovsky
4:30 P.M.	Western Fires Review	A. Willis, G. Hill, D. Livingstone
5:00 P.M.	Evening Free. We will have the Howard Johnson pretty much to ourselves and employees are encouraged to use this opportunity to swap tales, socialize and pick each others' brains. The lounge and hospitality rooms will be open and available.	

THURSDAY, DECEMBER 13

7:00 A.M.	Buffet breakfast or dine from menu at Howard Johnson Restaurant	
8:00 A.M.	General Session .... Cushnoc Auditorium, Temple Bowen	
8:00 A.M.	Safety Program and Record Review	Al Willis
8:30 A.M.	Bureau of Parks and Recreation	Lawrence Stuart
9:30 A.M.	Summary and What's Ahead	Fred E. Holt
10:00 A.M.	Group Meetings: Regional Safety Personnel	Al Willis
	Regional I&E Personnel	Walter Gooley
	New Movies and Slide Tape Programs	Temple Bowen

17. II. 50TH ANNIVERSARY OF THE MAINE FORESTRY DISTRICT 1909–1959

(Commemorative exercise for the second annual Forestry Field Day held at the University of Maine, Orono, including demonstrations of forest fire fighting equipment and 25-year service pin awards)

FOREWORD

The timberlands of Maine are recognized as its greatest natural resource. Maine owners were among the first in the country to recognize that forest fire protection was necessary. Prior to the Maine Forestry District Legislative Act of 1909 devastating fires occurred in the years 1899, 1903, and 1908. The area of unprotected wilderness of over 10 million acres had practically no fire protection system. Some landowners and operators were paying men to patrol their lands with their own funds. In 1903, the Legislature provided funds to assist in this work which was the start of an organized fire patrol system. The fires of 1908 showed the need for enlarging this system. To the landowners it seemed unjust to ask other taxpayers of the State for increased appropriations. Putting the burden where it belonged, they, together with other groups, decided to accept the entire burden of protection from forest fires through some form of an annual tax based on the dollar valuation. To legally accomplish this, and with the mutual consent of all concerned, the Maine Forestry District was enacted and incorporated fifty years ago in the year 1909. This cooperative approach to private forest land protection has served the best interests of the State of Maine, and the landowners, and will continue to provide this essential service.

18. KEEP MAINE GREEN SLOGANS *

The Keep Main Green program started in 1948, but the first record of slogans began in 1955. It was initiated by Joel Marsh, Supervisor of Information & Education of the Maine Forest Service.

Year	Slogan	Year	Slogan
1955	Keep Maine Green in 1955	1964	The Forests For Evermore: Keep Maine Green in '64
1956	Be Sure Our Program Clicks, Keep Maine Green in '56	1965	Keep Maine Green in 1965
1957	In 1957 So As You Say Keep Maine Green Every Day	1966	Fire and Forests Do Not Mix So Keep Maine Green in '66
1958	Let's Make It A Date Keep Maine Green in '58	1967	In '67 Do As You Say Keep Maine Green Every Day
1959	The Responsibility Is Yours and Mine To Keep Maine Green in 1959	1968	Let's Make a Date Keep Maine Green in '68
1960	Woodland Fires Are Very Risky So Keep Maine Green in '60	1969	The Responsibility is Yours and Mine To Keep Maine Green in '69
1961	Make Sure You are The One to Keep Maine Green in '61	1970	Now that '70 Is On The Scene Get Together and Keep Maine Green
1962	There is Something You Can Do, Keep Maine Green in '62	1971	Make Sure You Are The One To Keep Maine Green In '71
1963	We'll Have The Forests to Use & Sights to See, If We Keep Maine Green in '63	1972	There is Something You Can Do Keep Maine Green in '72

* Although these are state-wide fire prevention slogans, the message was spread to French-speaking pulp and lumber camps, sportsmen and fire personnel in the Maine Forestry District.

19. WOODS CLOSURE LAW AND AMENDMENTS

1) Chapter 52, Sections 1–4, 1909 Maine Governor has authority to suspend open season for hunting — prevent use of fire arms in forests during dangerous dry times — no reference to fishing.
2) Chapter 8, Sections 38–41, Revised Statutes, 1916 — Maine Same as original act of 1909.
3) Chapter 33, P.L. 1923, Maine Amending Chapter 8, Sections 38 and 39 of R.S. 1916 to include suspension of open season for fishing.
4) Chapter 11, Sections 38–41, R.S. 1930, Maine Same as Chapter 33, P.L. 1923 as amended — annulment by another proclamation.
5) Chapter 180, P.L. 1931 — Maine Can close sections of the state, also penalty on smoking or building of out of door fires.
6) Chapter 35, P.L. 1943 — Maine Amendment to payment of costs of enforcing provisions of proclamation.
7) Chapter 344, P.L. 1945 Lawful to build fires at Maine Forest Service Authorized Campsites during closure periods.
8) Chapter 36, Sections 105–108, R.S. 1954 — Maine
9) Title 12, Chapter 215, Sections 1151–1154, R.S. 1964 — Maine

20. WOODS CLOSURES - GOVERNOR'S PROCLAMATIONS
(Smoking, Building Out of Door Fires, Suspension of Open Season on Fishing and Hunting)

Date Issued	Date Effective	Date Annulled	No. of Days	Governor	Forest Commissioner
May 22, 1911	warning	—	—	Frederick W. Plaisted	Frank E. Mace
June 6, 1921	warning	—	—	Percival P. Baxter	Samuel T. Dana
Oct. 4, 1922	same day	Oct. 9	6	"	"
Oct. 10, 1923	warning	—	—	"	"
May 10, 1930	midnight 5/11/30	May 15	6	Wm. Tudor Gardiner	Neil L. Violette
Oct. 14, 1930	sundown 9/14/30	Oct. 16	3	"	"
May 19, 1932	sundown 5/19/32	May 27	9	"	"
June 1, 1934	sundown 6/1/34	June 13	13	Louis J. Brann	"
Aug. 19, 1935	sundown 8/19/35	Sept. 4	17	"	"
Aug. 31, 1937	immediately 8/31/37	Sept. 13	14	Lewis O. Barrows	"
Oct. 15, 1938[1]	immediately 9/15/38	Oct. 21	7	"	"
May 16, 1941[2]	sundown 5/16/41	July 25	12	Sumner Sewall	Raymond E. Rendall
Aug. 9, 1941[2]	same day 8/9/41	Aug. 18	10	"	"
Sept. 26, 1941[3]	same day 9/26/41	Oct. 4	9	"	"
June 1, 1944	sunrise 6/2/44	June 20	19	"	"
Aug. 15, 1944	sunrise 8/16/44	Sept. 5	21	"	"
Aug. 21, 1945[4]	sunrise 8/22/45	Sept. 5	15	Horace A. Hildreth	"
July 20, 1946[5]	sunrise 7/21/46	July 25	5	"	"
Oct. 16, 1947[5]	sunrise 10/17/47	Nov. 12	27	"	"
Oct. 21, 1947[6]	sunrise 10/22/47	Nov. 12	22	"	"
Sept. 9, 1948	sunrise 10/10/48	Oct. 12	2	"	A. D. Nutting
Sept. 30, 1948[7]	sunrise 10/1/48	Oct. 12	12	"	"
May 18, 1949	sunrise 5/19/49	May 25	7	Frederick G. Payne	"
June 15, 1949	sunrise 6/16/49	June 24	9	"	"
July 28, 1949	twelve noon 7/29/49	Sept. 1	35	"	"

302

July 18, 1952	midnight 7/18/52	Aug. 18	31	Frederick G. Payne	A.D. Nutting
July 10, 1953	midnight 7/10/53	July 14	4	Burton M. Cross	"
May 8, 1957	midnight 5/8/57	May 15	7	Edmund S. Muskie	"
Oct. 17, 1963[8]	midnight 10/18/63	Oct. 29	11	John H. Reed	Austin H. Wilkins
Oct. 18, 1963[9]	sundown 10/19/63	Oct. 29	10	"	"
Oct. 25, 1963[10]	sunrise 10/26/63	Oct. 29	3	"	"

---

[1] Oxford and Franklin Counties only – severe hurricane of 1938
[2] Somerset, Piscataquis and Penobscot Counties only
[3] Entire state closed except northern Aroostook County
[4] Closure extended for Hancock and Washington Counties to Aug. 28 (July 21–Aug. 28 – 39 days)
[5] Oct. 16 proclamation lifted Nov. 10 except York County and lifted Nov. 12
[6] Oct. 21 proclamation supercedes Oct. 16 proclamation to include suspension of open season for hunting
[7] Sept. 30 proclamation includes open season for hunting
[8] Oct. 17 proclamation prohibits smoking and building out-of-door fires only
[9] Oct. 18 proclamation suspension of open season for hunting
[10] Oct. 25 proclamation continued restriction to smoking and building out-of-door fires along a certain demarkation line of the state

Number of spring proclamations – 10; fall proclamations – 31

State of Maine
PROCLAMATION
By PERCIVAL P. BAXTER
Governor of Maine

Forest Fire Emergency

THE present forest fire situation in Maine, both within and without the Forestry District, is critical. As a result of a prolonged drought the woods are so dry that fires start and spread with dangerous rapidity. During the month of May there were over 250 fires—two-thirds as many as in the year 1920, itself an unusually bad year. These fires were due to carelessness, and many assumed alarming proportions. It required 1200 men to bring these fires under control, at an expense which already equals that of the entire year of 1920. Conditions are such that a conflagration is likely to occur which will sweep a large portion of the State causing a loss of life and property impossible to estimate.

NOW, THEREFORE, I, PERCIVAL P. BAXTER, Governor of Maine, do hereby proclaim that the present forest fire situation is an Emergency that should be brought to the attention of every citizen. I urge all persons, and particularly all campers, fishermen, and woodsmen, to use the utmost precautions. Every individual going into the Maine woods should do all in his power to prevent the starting of fires, should cooperate freely with the proper authorities in extinguishing fires, and should aid in bringing to justice those guilty of violating the forest fire laws. It is only by such cooperation that the State can pass through the present Emergency without suffering loss that will prove to be a calamity.

Given at the Office of the Governor at Augusta, and sealed with the Great Seal of the State of Maine, this sixth day of June, in the year of our Lord One Thousand Nine Hundred and Twenty-one, and of the Government of the State of Maine the One Hundred and First.

Percival P. Baxter,
Governor of Maine.

Attest:
Frank W. Ball,
Secretary of State.

STATE OF MAINE

1935 Sept
Lifted

PROCLAMATION

By the GOVERNOR

Suspension of the Open Season on Fishing

IN view of the serious forest fire menace existing in all parts of our State and upon recommendation of the Forest Commissioner and pursuant to the authority vested in me by virtue of Sections 38 to 41, inclusive, of Chapter 11 of the Revised Statutes as amended by Chapter 180 of the Public Laws of 1931, I do hereby proclaim

Suspension of the Open Season on Fishing

in the inland waters of the State, the same to be effective at sundown today and to continue until revoked by me.

This suspension applies to all sections of the State, and prohibits all smoking or the building of any and all fires out of doors, in the woods, but does not suspend or prohibit lawful fishing from boats or canoes on lakes, ponds, rivers or thoroughfares.

It is my earnest desire that all citizens of the State cooperate in this very serious situation and all game wardens and state officials are instructed to proceed in the enforcement of this proclamation.

Given at the Office of the Governor at Augusta and sealed with the Great Seal of the State of Maine, this nineteenth day of August in the year of our Lord one thousand nine hundred and thirty-five and in the one hundred fifth-ninth of the Independence of the United States of America.

[signature]
Governor

By the Governor:

Louis J. Brann
Secretary of State

State of Maine

1922 Oct 9
Lifted

PROCLAMATION BY THE GOVERNOR

Suspending Open Season for Hunting and Prohibiting the Carrying of Fire Arms in the Woods of Maine

The present forest fire situation in Maine is critical. As a result of the existing dry weather, fires in the woods once started spread with great rapidity. A sudden and alarming increase in the number of fires has occurred since the opening of the hunting season. The worst fire in the history of the State occurred in the month of October and several hundred thousand acres were burned over at that time. If present conditions continue disastrous conflagrations may result.

NOW, THEREFORE, I PERCIVAL P. BAXTER, Governor of the State of Maine, in accordance with the authority vested in me by Sections 38 to 41, Chapter 8 of the Revised Statutes, do hereby proclaim a suspension of the Open Season for hunting, said suspension to continue until revoked by me. All provisions of law covering and relating to the closed season shall continue in force during said period and persons violating the provisions of this Proclamation will be subject to the full penalty of the law.

In addition to the foregoing, whoever shoots during this period any wild animal or bird for the hunting of which there is no Closed Season, or whoever enters upon the wild lands of the State carrying, or having in possession, fire arms, will be punishable by a fine of One Hundred Dollars and costs.

This Proclamation does not prohibit the shooting of wild waterfowl on the tidal waters of the State.

I call upon all citizens of the State to comply with the provisions of this Proclamation and Hereby instruct all Game and Fire Wardens and all other State officials to enforce said provisions.

Given at the Office of the Governor at Augusta, and sealed with the Great Seal of the State of Maine, this Fourth Day of October, in the Year of our Lord One Thousand Nine Hundred and Twenty-two, and of the Government of the State of Maine the One Hundred and Third.

Percival P. Baxter,
Governor of Maine.

Attest:

Edgar C. Smith,
Deputy Secretary of State.

1931 sept 4th
Lifted oct 4th

STATE OF MAINE
PROCLAMATION
By the Governor

The continued lack of rain and drouth conditions within our State has resulted in a serious fire hazard in all parts of Maine. On the recommendation and request of the Forest Commissioner and pursuant to the authority vested in me by virtue of Sections 38-41 inclusive of Chapter 11 of the Revised Statutes, as amended by Chapter 180 of the Public Laws of 1931, I do proclaim a

Suspension of the Open Season on Fishing

in the inland waters of the State, the same to be effective immediately and to continue until revoked by me.

This suspension applies to all sections of the State and prohibits all smoking or the building of any and all fires out of doors in the woods, provided, however, that such suspension of open time shall not prohibit fishing from boats or canoes on ponds, lakes, rivers or thoroughfares.

We are very appreciative of the opportunity to entertain a record number of summer visitors this year within the borders of the State and I am naturally reluctant to take this extreme step, but in view of the unusually dry season and the desire to protect our forest resources which I feel is one of our great natural heritages, I am honoring the written appeal of the Forest Commissioner and shall be most anxious to remove the above suspension at the earliest possible moment.

It is my sincere desire that all citizens of the State shall understand and coöperate in this serious situation. All game wardens and State officials will be instructed to proceed in the enforcement of this proclamation.

Given at the office of the Governor at Augusta, and sealed with the Great Seal of the State of Maine, this thirty-first day of August, in the year of our Lord One Thousand Nine Hundred and Thirty-seven and of the Independence of the United States of America the One Hundred and Sixty-second.

Louis J. Brann
Governor

By the Governor:
Frederick Robie
Secretary of State

1941 oct 4th
Lifted

State of Maine
Proclamation by the Governor

The continuing lack of rain and the unusual drought conditions in the forests throughout the State, with the exception of northern Aroostook County, has resulted in an extremely serious fire hazard.

On the recommendation and request of the Forest Commissioner and pursuant to the authority vested in me by virtue of Sections 38 to 41, inclusive, of Chapter 11 of the Revised Statutes as amended by Chapter 180 of the Public Laws of 1931, I hereby proclaim suspension of the open season for hunting and fishing and I do hereby proclaim a

PROHIBITION OF SMOKING AND THE BUILDING OF FIRES

out of doors, in the woods, throughout the State except that portion of Aroostook County lying to the north of the borders of Somerset, Piscataquis, and Penobscot Counties and north of Township 8, Range 5, Township 8, Range 4, Township 8, Range 3, Township C, Range 2, and Monticello in the County of Aroostook. Such suspension and prohibition shall continue until such hazard ceases.

This suspension and prohibition does not suspend or prohibit lawful fishing and hunting from boats or canoes on lakes, ponds, rivers, or thoroughfares.

It is my earnest desire that all citizens of the State cooperate in an effort to prevent the occurrence of fires by exercising particular caution to prevent violation of this proclamation on smoking and the lighting of fires.

All fire wardens, all game wardens, and all state officials are instructed to proceed forthwith in the enforcement of this proclamation.

Given at the Office of the Governor at Augusta and sealed with the Great Seal of the State of Maine, this twenty-sixth day of September in the year of our Lord one thousand nine hundred and forty-one and in the one hundred sixty-fifth of the Independence of the United States of America.

STATE OF MAINE

Proclamation

BY THE GOVERNOR

In view of the extremely dry condition and serious fire hazard existing in the forests of Maine and upon recommendation of the State Forest Commissioner and pursuant to the authority vested in me by virtue of Sections 38 to 41, inclusive, of Chapter 11 of the Revised Statutes as amended by Chapter 180 of the Public Laws of 1931, I do hereby proclaim suspension of the open season for fishing in the inland waters of the State, except lakes, ponds, rivers, or thoroughfares when fishing from boats or canoes.

This proclamation prohibits all smoking or the building of any and all fires out of doors in the woods. It does not suspend or prohibit lawful fishing from boats or canoes on lakes, ponds, rivers and thoroughfares.

This proclamation shall be effective at sunrise August 16, 1944, and shall continue until annulled by further proclamation.

It is essential that all persons in the State cooperate fully in this serious situation. All fire wardens and all game wardens and state officials shall proceed in the enforcement of this proclamation.

Given at the Office of the Governor at Augusta and sealed with the Great Seal of the State of Maine, this fifteenth day of August in the year of our Lord One Thousand Nine Hundred and Forty-four, and of the Independence of the United States of America, the One Hundred and Sixty-eighth.

State of Maine

Proclamation

By The Governor

Prolonged drought conditions with weather predictions indicating no relief in sight have resulted in a serious fire hazard to the forests of Maine. On the recommendation of the State Forest Commissioner and pursuant to the authority vested in me by virtue of Chapter 344 of the Public Laws of 1945, I do hereby suspend the open season for hunting in all sections of the forests of the State and I do hereby prohibit all smoking or the building of any and all fires out of doors in the woods except at public camp sites maintained by the Forestry Department.

This proclamation shall be effective at sunrise, October 17, 1947, and shall continue until annulled by further proclamation.

It is essential that all persons in the State cooperate fully in this serious situation. All fire wardens and all game wardens and state officials shall proceed in the enforcement of this proclamation.

Given at the Office of the Governor at Augusta and sealed with the Great Seal of the State of Maine, this sixteenth day of October, in the year of our Lord One Thousand Nine Hundred and Forty-seven, and of the Independence of the United States of America, the One Hundred and Seventy-second.

Governor

By the Governor

Secretary of State

A true copy.

Attest: HAROLD I. GOSS,
Secretary of State

State of Maine
Proclamation
By The Governor

Whereas, many raging fires are being reported in the woods in various parts of the State, and

Whereas, the long and continued drought has increased the menace of fire in the forests of the State, and

Whereas, it appears to me that hunting is likely to increase such menace and fire hazard,

Now, Therefore, I, Horace Hildreth, Governor of the State of Maine, pursuant to the authority vested in me by Chapter 344 of the Public Laws of 1945, do hereby suspend the open season for hunting, and hunting is hereby prohibited in all parts of the State, and I also prohibit all smoking or the building of fires out of doors in all sections of the woods in the State.

This proclamation shall be effective at sunrise, October 22, 1947, and shall continue until annulled by me, by further proclamation.

This proclamation is to supersede the proclamation I heretofore issued, on October 16, 1947, which is hereby annulled.

I hereby urgently request all persons in the State to lend their fullest cooperation in the observance of this proclamation.

All fire wardens, game wardens and all peace officers of the State are hereby ordered to enforce the provisions of this proclamation.

Given at the Office of the Governor at Augusta and sealed with the Great Seal of the State of Maine, this twenty-first day of October, in the year of our Lord One Thousand Nine Hundred and Forty-seven, and of the Independence of the United States of America, the One Hundred and Seventy-second.

Horace Hildreth
Governor

By the Governor

State of Maine
Proclamation
By the Governor

Prolonged drought, allowing forest fires to burn deeply into the ground, has resulted in a serious fire hazard to the forests of Maine. Twenty forest fires are now under patrol. Weather predictions indicate no relief in sight. On the recommendation of the State Forest Commissioner and pursuant to the authority vested in me by virtue of Chapter 344 of the Public Laws of 1945, I do hereby prohibit all smoking or the building of any and all fires out of doors in the woods, except at public camp sites maintained by the Forestry Department. I urge sportsmen to cooperate with us in enforcing the campfire and smoking ban when fishing in any and all Maine waters.

This proclamation shall be effective at twelve noon, Friday, July 29, 1949, and shall continue until annulled by further proclamation.

It is essential that all persons in the state cooperate fully in this serious situation. All fire wardens, all game wardens, and state officials shall proceed in the enforcement of this proclamation.

Given at the Office of the Governor at Augusta, and sealed with the Great Seal of the State of Maine, this twenty-eighth day of July, in the year of our Lord One Thousand Nine Hundred and Forty-nine, and of the Independence of the United States of America, the One Hundred and Seventy-fourth.

By the Governor

Harold I. Goss
Secretary of State

Governor

A true copy.

Attest: HAROLD I. GOSS,
Secretary of State

BIBLIOGRAPHY

(Sources Used in Preparation of This Manuscript)

I. DOCUMENTS:

Elmer Crowley, "Diary." (Dept. of Conservation)

Files of M.F.D. Advisory Committee meetings. (Dept. of Conservation)

Forest Commissioners' Biennial Reports, 1891–1972 (Maine State Library): 1894; 1896, pp. 20–23; 1903–04 (on forest fires); 1903; 1909; 1910; 1912; 1917–18; 1927; 1947 and 1972 (on petty cash funds, financing and accounting); 1905–06; 1909–10; 1911–12 (on the creation of the M.F.D. and the advisory committee); 1905–20; 1943–44; 1967–1972 (on conversion from lookout towers to aircraft); 1909; 1927–28 (on organization and personnel of the M.F.D.); 1921–22; 1945–46; 1949–50 (on financing spruce budworm control); 1933–34, pp. 112–19; 1935–36, pp. 113–28; 1937–38, pp. 127–37; 1939–40, pp. 64–69; 1941–42, p. 68 (on M.F.D. involvement with the CCC); 1941–46 (on M.F.D. involvement in WW II); 1943–46; 1949–54; 1957–58; 1971–72 (on conversion from woods telephone to radio); 1957–58, pp. 91–93; 1959–60, pp. 126–31 (on forestry field days); 1960–72 (M.F.D. budget reports for advisory committee consideration and action); 1971–72 (on equipment and labor rates changes in forest fire suppression; capital valuation of real estate and contents); August 1972 (to the 106th Maine Legislative Research Subcommittee on Forest District Taxation).

Forestry Department (now Department of Conservation, State House, Augusta):

Annual insurance schedules and personnel directories (1903–1972)

Bulletins: see especially #9 (1915) on forest fire causes

Data on nomenclature in the warden service (Personnel file case)

Petty cash accounting records

Photo Album

Records of the Northeastern Fire Protection Commission

Reports: 1963 (on Public Reserved Lots); 1973 (on labor and equipment rates); Entomology Division, 1967, 1969, 1970, 1971, 1972 (on budworm control)

Retro-Respective Rating Plan Policy

Scrapbooks

The John Weeks Law Story (U.S. Dept. of Agriculture, Forest Service, Washington, D.C.)

Land Agents' Reports 1820–1891 (Dept. of Conservation, State House, Augusta)

Land Office Records (Maine State Archives), 1975

Legislative Committee on Natural Resources Report (January 1975) on its study of the spruce budworm problems in Maine (Dept. of Conservation)

"Lightning in Relation to Forest Fires" (Washington, D.C.: U.S. Forest Service Bulletin #111, 1912), pp. 12, 80

Maine Environmental Impact Statement for 1973 to the Environmental Protection Agency, Washington, D.C. (Entomology Division, Dept. of Conservation): p. 2, and Appendix C, pp. 4–6

Maine House Journal: 1891, p. 170; 1909, pp. 63, 225, 443, 738, 748, 786, 862, 900, 988, 1044 and 1048

Maine Legislative Record (State Law Library, Augusta): 1909, re the creation of the M.F.D.

Maine Public Laws (State Law Library, Augusta): Chapters 100 (1891: Sections 1–18); 168 (1909: Section 1, Salary increase of forest commissioner); Title 12, on Taxation (1964: Sections 1201–03; 1301–02; 1601; 1601A–1607, M.R.S.A.); Chapters 441 (1965: Creation of State Archives); 476 (Section 15, R.S.T. 12, Section 501, amended 1967– Salary change to 100 percent General Fund appropriation); 572 (1968, Salary increase for forest commissioner); 147 (1969: Unclassified to classified service); 531 (1971: Salary increase of forest commissioner); 87 (1973: Fire protection, Baxter State Park); 141 (1973: Fire protection, Indian Township); 460 (1973: Creation of Department of Conservation)

Maine Municipal Association publications

The Maine Register (Portland: Fred L. Tower Co.) 1973–74

Maine Senate Journal, 1891: pp. 150, 263, 557, 597, 1909: pp. 381, 414, 759, 781, 821, 933, 942

W. B. Salkald, Canadian International Paper Co.: "Report Covering Forest Fires in Maine Townships, Daaquam Division" (International Paper Co., Jay, Maine) 1934

Proctor Report on Maine Indians (Maine Legislative Research Committee Study), 1942. (State Law Library)

Records of former Chief Warden John Mtichell (on file at Dept. of Conservation)

Records at Radio Repair Shop, Bolton Hill, Augusta

Records: *State Personnel Department*

"Report on the Public Lots," Lee Schepps, Asst. Attorney General (Sept. 12, 1972)

Reports State Tax Assessor's Office, (State House, Augusta) on the creation of the M.F.D. and advisory committee; land ownership; state valuation procedures (January 1969)

Timber Resources of Maine (10-year periodic study put out by the Department of Conservation, State House, Augusta): 1971, pp. 8 and 35

Austin H. Wilkins, "Maine Timberland Ownerships," paper for the Legislative Research Committee 1970 (State Law Library)

Woods Closure: legislation and proclamations: see 1) Documents Division, Secretary of State's Office; 2) Bureau of Forestry (Augusta) file of proclamations; 3) Statutory references on the law and amendments (State Law Library)

II. BOOKS

Philip T. Coolidge, *History of the Maine Woods* (Bangor, Me.: Furbush Roberts Printing Co., 1963), pp. 191–82.

Forest Watchman's Handbook (Forest Fire Control Series #1), pp. 4–5. (Forestry Service, Augusta, Maine Department of Conservation)

Stewart H. Holbrook, *Burning an Empire* (New York: The Macmillan Co., 1943), pp. 54–59

Yankee Logger (New York: International Paper Co., 1961).

Henry David Thoreau, *The Maine Woods*. (New York: Bramhall House, 1950), p. 289

Ralph R. Widner, Ed., *Forest and Forestry in the American States*. Compiled by the National Association of State Foresters (Washington, D.C., Dept. of Agriculture, 1968), pp. 13–20; 120–23

About the Author

AUSTIN H. WILKINS was born in Massachusetts in 1903. He was graduated from the University of Maine, Orono, Maine in 1926 with a B.S. degree in forestry, and from Cornell University, Ithaca, New York, in 1927 with an M.F. degree. He studied forestry in Europe in 1927 under Dr. C. A. Schenck, internationally known German forester. In 1931 he was assigned as instructor to augment the University of Maine forestry faculty at the annual winter camp, Princeton, Maine.

He served forty-four continuous years with the State of Maine Forestry Department in Augusta from 1928 through 1972, and was forest commissioner for the last fourteen years. During his tenure of public office he served under thirteen governors.

He is a past president of the National Association of State Foresters (1965), past chairman of the New England Section of the Society of American Foresters (1946–47), and past chairman of the Baxter State Park Authority (1959–71).

He has received several state, regional, and national forestry awards and citations. He has written articles for numerous periodicals and magazines and presented papers at forestry conferences, besides writing *The Forests of Maine Bulletin #8, 1932,* for the State Forestry Department.

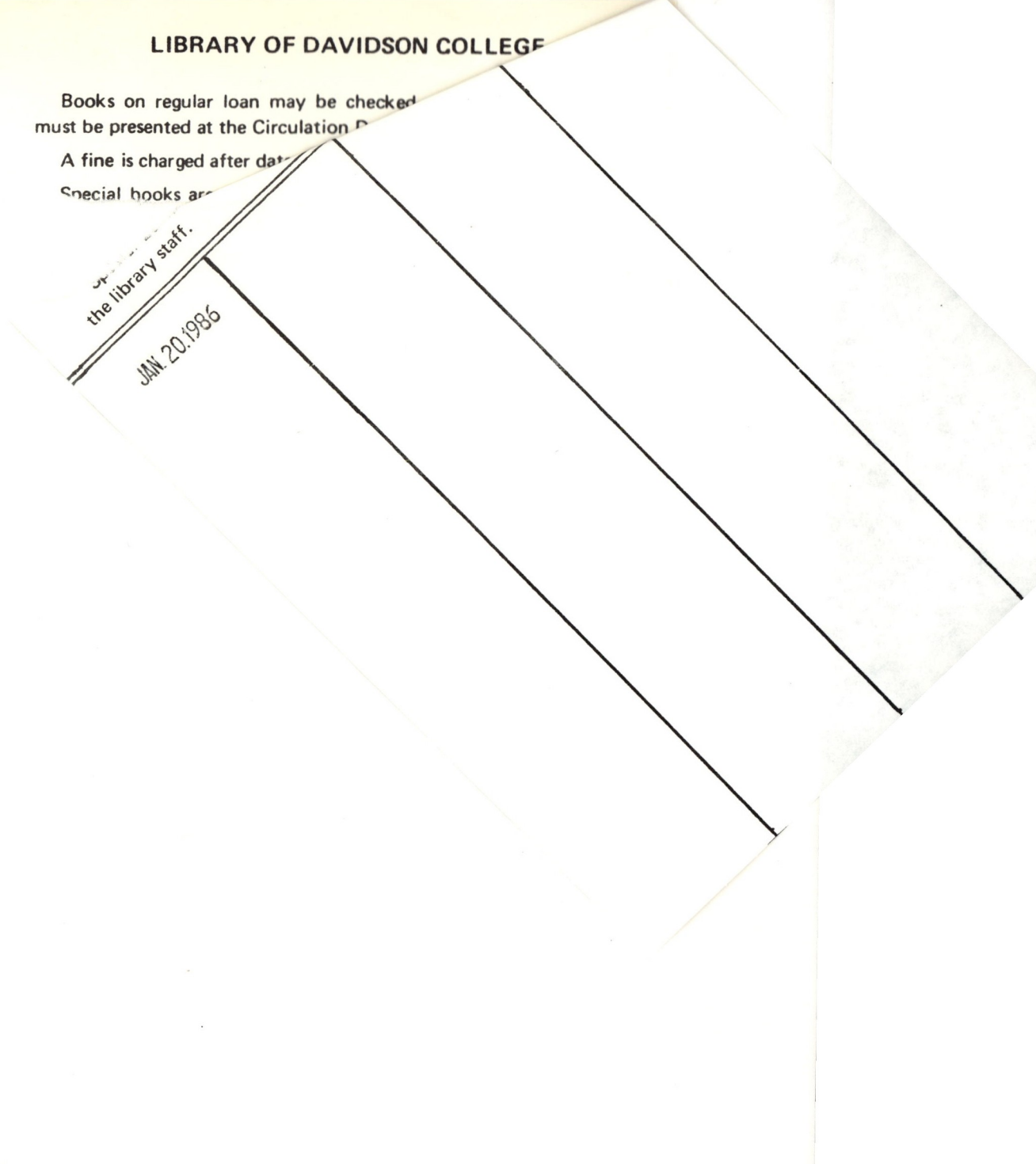